Laboratory Control System Operations in a GMP Environment

Laboratory Control System Operations in a GMP Environment

David M. Bliesner, Ph.D.
Delphi Analytical Services, Inc.
Indian Rocks Beach, Florida

WILEY

This edition first published 2020

Registered Office
John Wiley & Sons, Inc., 111 River Street, Hoboken, NJ 07030, USA

Editorial Office
111 River Street, Hoboken, NJ 07030, USA

For details of our global editorial offices, customer services, and more information about Wiley products visit us at www.wiley.com.

Wiley also publishes its books in a variety of electronic formats and by print-on-demand. Some content that appears in standard print versions of this book may not be available in other formats.

Library of Congress Cataloging-in-Publication Data

Names: Bliesner, David M., author.
Title: Laboratory control system operations in a GMP environment / David M. Bliesner.
Description: Hoboken, NJ : Wiley, [2020] | Includes bibliographical references and index.
Identifiers: LCCN 2019052264 (print) | LCCN 2019052265 (ebook) | ISBN 9781119529231 (cloth) | ISBN 9781119529279 (adobe pdf) | ISBN 9781119529293 (epub)
Subjects: LCSH: Medical laboratories–Quality control. | Laboratories–Management. | Pharmacy–Research. | Manufacturing processes–Quality control.
Classification: LCC RB36.3.Q34 B55 2020 (print) | LCC RB36.3.Q34 (ebook) | DDC 610.28/4–dc23
LC record available at https://lccn.loc.gov/2019052264
LC ebook record available at https://lccn.loc.gov/2019052265

Cover Design: Wiley
Cover Image: © ImageJournal-Photography/Getty Images

Set in 9.5/12.5pt STIXTwoText by SPi Global, Chennai, India

Printed in the United States of America

10 9 8 7 6 5 4 3 2 1

To my wife and best friend, Kathy.

Contents

Preface

Regulatory agencies such as the US Food and Drug Administration (FDA) and the European Medicines Agency (EMA) mandate that companies who manufacture drugs for human and animal use be operated in a state of control by employing conditions and controls commonly referred to as good manufacturing practices (GMPs). In the United States the GMPs or CGMPs (where C = current) are codified in 21 Code of Federal Regulations Parts 210 and 211. In the European Union the GMPs are described in EudraLex – Volume 4, the Good Manufacturing Practice (GMP) guidelines. Other Regulatory bodies, such as the World Health Organization (WHO) and the International Council for Harmonization (ICH) also provide requirements and guidelines with respect to GMPs.

At the turn of the twenty-first century the FDA was actively engaged in regulatory actions against pharmaceutical companies whom they determined to be in significant violation of the CGMPs. As part of FDA's actions, several large companies entered into voluntary legal agreements referred to as Consent Decrees. These agreements existed for the expressed purpose of correcting the deficiencies related to CGMPs. Consent Decrees routinely require the contracting of independent third-party expert consultants. These consultants' duties include performing a baseline audit of the offending facilities, documenting their findings and helping the companies develop and implement corrective action plans.

To help collect evidence to support regulatory actions, FDA began using a Quality Systems approach to evaluate a firm's level of GMP compliance.

The Quality Systems approach is a scheme of systems for the manufacture of drugs and/or drug products. The general scheme of systems FDA used for auditing manufacture facilities consists of the following:

1. Quality System
2. Facilities and Equipment System
3. Materials System
4. Production System
5. Packaging and Labeling System
6. Laboratory Control System

According to FDA "The Quality System provides the foundation for the manufacturing systems that are linked and function within it."[1]

This approach is commonly referred to as the six-system model and is still used today by the FDA to conduct inspections of GMP facilities.

During this time, Delphi Analytical Services, Inc. (Delphi) served as one of the third-party expert consulting firms on several Consent Decrees at major pharmaceutical companies. Delphi's core competency is the practical understanding of workings of the laboratory control system (LCS), the sixth quality system mentioned in the FDA model. Delphi was extensively involved in performing third-party audits of quality control (QC) and research and development (R&D) laboratories for companies under Consent Decree.

While executing these consulting assignments it became apparent that the CGMPs regulations, FDA Guidance Documents and FDA Internal Compliance Program Guides offered little direction or specifics of what was expected of a CGMP compliant LCS. Because of this, Delphi developed and documented a systematic means for performing the baseline audits, capturing deficiencies, reporting the results, and developing subsequent corrective action plans, for QC laboratories. In addition, Delphi also created instructional materials and began teaching courses on auditing QC laboratories at various client sites and scientific conferences.

Due to the inherent complexity of the LCS, Delphi divided it into seven sub systems or sub elements. These included the following:

1. Laboratory Managerial and Administrative Systems
2. Laboratory Documentation Practices and Standard Operating Procedures
3. Laboratory Equipment Qualification and Calibration
4. Laboratory Facilities
5. Methods Validation and Technology Transfer
6. Laboratory Computer Systems
7. Laboratory Investigations

Based on real-world experiences, Delphi then, in turn, developed detailed audit checklists for each LCS sub element. These became the basis for most of Delphi's consulting contracts and instruction material.

In March 2004 at the Pittsburgh Conference on Analytical Chemistry and Applied Spectroscopy (PittCon©), Delphi Analytical Services, Inc.'s president, Dr. David M. Bliesner, Ph.D., taught a one-day course titled "How to Establish a GMP Laboratory Audit System." The course was well attended which led to a meeting with a representative of Wiley-Interscience, John Wiley & Sons, Inc. At the conclusion of that meeting, it was agreed that he would write a proposal

1 FDA Guidance for Industry Quality Systems Approach to Pharmaceutical CGMP Regulations, September, 2006.

for publication of a book. The proposal was accepted and became the book titled *Establishing a CGMP Laboratory Audit System: A Practical Guide* in 2006.

Over the course of the next 12 years (2006–2018), the pharmaceutical industry experienced a massive paradigm shift in its business and operational models. This in turn resulted in changes to the manner in which regulatory agencies enforced the GMPs. Some key elements of these changes included:

- Substantial outsourcing of drug development, manufacturing, testing, and other functions
- Offshoring of manufacturing and testing of active pharmaceutical ingredients (APIs) and drug products
- Profit versus quality driven decision making
- Reductions in experienced, permanent workforces
- Increased regulatory activity by US FDA at overseas plants
- Increased Global Regulatory Agency oversight, guidance, and cooperation

Because of these changes, and other factors, it became obvious that *Establishing a CGMP Laboratory Audit System: A Practical Guide* had much broader utility than just serving as an audit guide. In particular, the real value of the book over the years has not been as an auditing guide but:

- A means to introduce auality assurance (QA) and QC personnel to the concept of quality systems and the LCS in particular
- As a basis for instructing QA and QC personnel on Laboratory GMPs
- To assist in interpreting and clarifying regulatory expectations for the LCS
- As the basis for assisting laboratory management in implementing and maintaining CGMP compliant practices in their own QC laboratories

Therefore, the purpose of this new book titled *Laboratory Control System Operations in a GMP Environment* is to focus on LCS *operations* versus *auditing*.

The goal in writing this book is to publish a text, which when implemented in QC and R&D laboratories, provides the basis for operating a CGMP compliant LCS. This will improve an organization's chances for withstanding regulatory scrutiny and enhance operational efficiency. This new book is designed to be used in several ways, including:

- Designing and implementing a new, from scratch, CGMP compliant LCS
- Upgrading or tweaking an existing LCS
- Laying a basis for initial and periodic LCS GMP training
- Laying a basis for improving operational efficiency
- Serving as an operational reference guide: Third party "sanity" check to help solve compliance challenges as they arise

Laboratory Control System Operations in a GMP Environment builds on the original auditing text described earlier. However, instead of emphasizing auditing, this new text not only focuses and upgrades the operational aspects of the original seven LCS sub elements but also expands the LCS to encompass three additional sub elements, which are marked with an asterisk(*) in the following list. This results in the following 10 chapters in the book:

1. Laboratory Managerial and Administrative Systems
2. Laboratory Documentation Practices and Standard Operating Procedures
3. Laboratory Equipment
4. Laboratory Facilities
5. Method Validation and Method Transfer
6. Laboratory Computer Systems
7. Laboratory Investigations
8. **Data Governance and Data Integrity***
9. **Stability Program***
10. **General Laboratory Compliance Practices***

Each chapter in this text describes the critical functions of the LCS sub element so the reader understands what is expected from the FDA and other Global Regulatory Agencies. In addition, each chapter links to tools, templates, checklists, and Global Regulatory Agencies' guidances. All of these tools and templates are accessible for download online through a Wiley Interscience web portal for easy modification and application by the end-user in their own laboratories.

Readers, in using this book, may accomplish the following:

- Fully implement a functional LCS which can withstand Global Regulatory scrutiny
- Increase operational efficiency
- Stay current with GMPs and industry trends
- Save time by using the real-world tools and templates found in the book, which can be modified and used by the reader
- Use the text as a benchmark reference to which they can assess the status of compliance of their own laboratories

Laboratory Control System Operations in a GMP Environment is written for a broad audience. It is applicable to both QC and QA professionals in small, medium, and large companies within the pharmaceutical and biopharmaceutical industries. R&D personnel working in non-GMP environments will also benefit by applying the organizational schemes and principals presented in this text.

This book is particularly helpful for personnel who work in smaller companies because they often do not have the financial, personnel resources, and existing "corporate knowledge" that large US- and European-based companies may possess. This means that smaller organizations are often left to "figure it out" on their own. For these smaller operations, this text is particularly valuable because of the example-templates and checklists it includes.

To our knowledge no such detailed operational text or guide exists in the marketplace. We hope you find *Laboratory Control System Operations in a GMP Environment* useful and wish you the best in your continuing quest to establish a quality-minded culture, improve operational efficiency, and thrive under Global Regulatory scrutiny.

January 2020

David M. Bliesner, Ph.D.
Indian Rocks Beach, Florida

About the Companion Website

This book is accompanied by a companion website:

www.wiley.com/go/Bliesner/LabControl_GMPEnvironment

Scan this QR code to visit the companion website

The website includes Appendices and Weblinks.

1

Introduction to the Quality Systems Based Approach to CGMP Compliance

Overview of Quality Systems and the Laboratory Control System

The US Food and Drug Administration (US FDA) mandates that a drug firm, and therefore the laboratory, be operated in a state of control by employing conditions and practices that assure compliance with the intent of the Federal Food, Drug, and Cosmetic Act and portions of the Current Good Manufacturing Practice (CGMP) regulations (e.g. 21 CFR Parts 210 and 211) that pertain to it. Activities found in drug firms, including operation of the laboratory, can be organized into systems that are sets of operations and related activities. Control of all systems helps to ensure the firm will produce drugs that are safe, have the proper identity and strength, and meet the quality and purity characteristics as intended [1, 2].

For drug firms, FDA has outlined the following general scheme of systems that impact the manufacture of drugs and drug products:

(1) *Quality System*. This system assures overall compliance with CGMPs and internal procedures and specifications. The system includes the quality control (QC) unit and all of its review and approval duties (e.g. change control, reprocessing, batch release, annual record review, validation protocols, reports, etc.). It also includes all product defect evaluations and evaluation of returned and salvaged drug products. (See the CGMP regulation, 21 CFR 211 subparts B, E, F, G, I, J, and K.)
(2) *Facilities and Equipment System*. This system includes the measures and activities that provide an appropriate physical environment and resources used in the production of the drugs or drug products. It includes:
   (a) Buildings and facilities along with maintenance.
   (b) Equipment qualifications (installation and operation); equipment calibration and preventative maintenance; and cleaning and validation of cleaning processes as appropriate. Process performance qualifications are

Laboratory Control System Operations in a GMP Environment, First Edition. David M. Bliesner.

Companion website: www.wiley.com/go/Bliesner/LabControl_GMPEnvironment

included as part of process validation, which is done within the system where the process is employed.

(c) Utilities that are not intended to be incorporated into the product such as heating, ventilation, and air-conditioning (HVAC), compressed gases, steam, and water systems. (See the CGMP regulation, 21 CFR 211 subparts B, C, D, and J.)

(3) *Materials System.* This system includes measures and activities to control finished products and components including water or gases that are incorporated into the product, containers, and closures. It includes validation of computerized inventory control processes, drug storage, distribution controls, and records. (See the CGMP regulation, 21 CFR 211 subparts B, E, H, and J.)

(4) *Production System.* This system includes measures and activities to control the manufacture of drugs and drug products including batch compounding, dosage form production, in-process sampling and testing, and process validation. It also includes establishing, following, and documenting performance of approved manufacturing procedures. (See the CGMP regulation, 21 CFR 211 subparts B, F, and J.)

(5) *Packaging and Labeling System.* This system includes measures and activities that control the packaging and labeling of drugs and drug products. It includes written procedures, label examination and usage, label storage and issuance, packaging and labeling operations controls, and validation of these operations. (See the CGMP regulation, 21 CFR 211 subparts B, G, and J.)

(6) *Laboratory Control System.* This system includes measures and activities related to laboratory procedures, testing, analytical method development, validation and/or qualification/verification, and the stability program. (See the CGMP regulation, 21 CFR 211 subparts B, I, J, and K.)

According to FDA, "The Quality System provides the foundation for the manufacturing systems that are linked and function within it." Graphically, the scheme of systems looks like [1] (Figure 1.1):

This approach is commonly referred to as the six-system model and is still used today by FDA to conduct inspections of good manufacturing practice (GMP) facilities.

As stated in (6) earlier, FDA considers a firm's Laboratory Control System (LCS) to be a key element in CGMP compliance. Within the LCS are at least 10 additional sub systems or sub elements, which may include:

- Laboratory Managerial and Administrative Systems (MS)
- Laboratory Documentation Practices and Standard Operating Procedures (OP)
- Laboratory Equipment (LE)
- Laboratory Facilities (LF)
- Method Validation and Method Transfer (MV)

Figure 1.1 FDA guidance for Industry Quality Systems approach to pharmaceutical CGMP regulations, September 2006 [1].

- Laboratory Computer Systems (LC)
- Laboratory Investigations (LI)
- Data Governance and Data Integrity (DI)
- Stability Program (SB)
- General Laboratory Compliance Practices (CP)

These 10 sub elements of the LCS are not part of any guideline document, international council, or inspection convention. Instead they have been created by the author, to promote the establishment and maintenance of Quality Systems and sub systems, which demonstrate you are in control of your laboratory operations and thus in compliance with the CGMP regulations.

These 10 sub element topics constitute 10 chapters within this book. Each chapter will describe the critical functions of the LCS sub element so the reader understands what is expected from the US FDA and other Global Regulatory Agencies.

A listing of the primary Global Regulations, the Agencies that enforce them, and the international councils or inspection conventions that help to harmonize their efforts are listed in the succeeding text.

Regulations and Regulatory Bodies

The primary, globally significant, regulations related to the manufacturing, processing, packing, or holding of drugs include:

- 21 Code of US Federal Regulations Part 210 and 211 Current Good Manufacturing Practice Regulations
- EudraLex – Volume 4 – Good Manufacturing Practice (GMP) guidelines

The major regulatory bodies or organizations that enforce the regulations or assist in harmonizing international regulatory efforts include:

- US Food and Drug Administration (US FDA, United States)
- European Medicines Agency (EMA, European Union)
- Medicines and Healthcare products Regulatory Agency (MHRA, United Kingdom)
- Health Canada (Canada)
- Brazilian Health Regulatory Agency (ANVISA, Brazil)
- Pharmaceuticals and Medical Devices Agency (PMDA, Japan)
- Therapeutic Goods Administration (TGA, Australia)
- World Health Organization (WHO-International)
- Central Drugs Standard Control Organization (CDSCO, India)
- The International Council for Harmonisation of Technical Requirements for Pharmaceuticals for Human Use (ICH-International)
- Pharmaceutical Inspection Convention (PIC) and the Pharmaceutical Inspection Co-operation Scheme (PIC Scheme) (PIC/S-International)

There are numerous other country-specific bodies, which enforce their own laws related to the manufacturing, processing, packing, or holding of drugs. The reader is encouraged to consult the requirements of their own country's laws and regulations regarding the manufacture of pharmaceuticals.

Regulatory Guidance

Traditionally, Regulatory Agencies themselves have provided limited insight and assistance into how organizations operating within the pharmaceutical industry can comply with the regulations. However, over time, regulatory guidances and other instruments have arisen and evolved and today consist of a fairly large body of knowledge, which can be used by organizations to aid in compliance with the CGMPs.

When it comes to regulatory guidance for Quality Control (QC) Laboratories, the following documents may be helpful:

- US FDA Compliance Programs to FDA staff, Chapter 56: Drug Quality Assurance 7366.002 Drug Manufacturing Inspections
- US FDA Guidance for Industry, Quality Systems Approach to Pharmaceutical CGMP Regulations
- ICH Harmonised Tripartite Guideline, Q1A to Q1F Stability
- ICH Harmonised Tripartite Guideline, Q2 Analytical Validation
- ICH Harmonised Tripartite Guideline, Q3A to Q3D Impurities
- ICH Harmonised Tripartite Guideline, Q4 to Q4B Pharmacopoeias
- ICH Harmonised Tripartite Guideline, Q6A to Q6B Specifications
- ICH Harmonised Tripartite Guideline, Q7 Good Manufacturing Practice Guide for Active Pharmaceutical Ingredients
- ICH Harmonised Tripartite Guideline, Q8 Pharmaceutical Development
- ICH Harmonised Tripartite Guideline, Q9 Quality Risk Management
- ICH Harmonised Tripartite Guideline, Q10 Pharmaceutical Quality System
- ICH Harmonised Tripartite Guideline, Q12 Lifecycle Management
- ICH Harmonised Tripartite Guideline, Q14 Analytical Procedure Development
- WHO Annex 2: Good Manufacturing Practices for Pharmaceutical Products: Main Principles
- FDA Guidance for Industry Quality Systems Approach to Pharmaceutical CGMP Regulations, September 2006

It should be noted that although not legally binding, violation of the principals of ICH Harmonised Tripartite Guideline, Good Manufacturing Practice Guide for Active Pharmaceutical Ingredients, Q7, are sometimes documented as findings by FDA.

Additional FDA and ICH guidelines exist and can be located at: https://www.fda.gov/drugs/guidances-drugs/all-guidances-drugs, https://www.ich.org/products/guidelines/quality/article/quality-guidelines.html, and https://www.fda.gov/drugs/guidance-compliance-regulatory-information/drug-compliance-programs

Application of This Text

The remainder of "Laboratory Control System Operations in a GMP Environment" is dedicated to describing the critical functions of the LCS sub elements so the reader understands what is expected from the FDA and the Global Regulatory Agencies listed earlier. In addition, each chapter will present or link to tools, templates, checklists, and some of the Global Regulatory Agencies' guidance listed previously.

It should be noted that text is written for a broad audience. It is applicable to both Quality Control and Quality Assurance professionals in small, medium, and large companies within the pharmaceutical and biopharmaceutical industries. R&D personnel working in non-GMP environments will also benefit applying the organizational schemes and principals presented in this text.[1] Also, forcign firms in China and India will find this book especially useful.

This book is particularly helpful for personnel who work in smaller companies because they often do not have the financial, personnel resources, and existing "corporate knowledge" that a large US- and European-based company may have and are therefore often left to "figure it out" on their own. In this respect the guide is particularly valuable in the example-templates and checklists it includes.

Overlap and Redundancy

As the reader progresses through this text, they will notice that some topics, notes, and clarifications are addressed more than once and in different locations within the book. This was done by the author on purpose to ensure that important topics are addressed appropriately and reinforced.

Additionally, the QC laboratory is a very complex and dynamic entity, which continually grows and evolves over time. This means that the 10 sub elements into with the LCS is divided (which is purely a matter of choice on the part of the author) can be reduced, modified, or expanded to address changes within the organization and the evolution of Regulatory Agency expectations and standard industry practices. This is why there is a "C" in CGMP: C means current, which is today, not yesterday.

Tools and Templates

The following are provided in electronic format in the Chapter 1 Appendix (www.wiley.com/go/Bliesner/LabControl_GMPEnvironment):

- 21 Code of Federal Regulations Parts 210 and 211 – Current Good Manufacturing Practice Regulations, Revised as of April 1, 2005
- US FDA Compliance Programs to FDA staff, Chapter 56: Drug Quality Assurance 7366.002 Drug Manufacturing Inspections, October 31, 2017.
- US FDA Guidance for Industry, Quality Systems Approach to Pharmaceutical CGMP Regulations, September 2006.

---

1 In June 2018 the ICH Assembly endorsed ICH Q14 "Analytical Procedure Development Guideline" and thus, by default, requiring a level of compliance with the GMPs by R&D laboratories.

References

1 21 CFR Parts 210 and 211 Current Good Manufacturing Practice for Finished Pharmaceuticals.
2 US FDA (2017). Compliance Programs to FDA staff, Chapter 56: Drug Quality Assurance 7356.002 Drug Manufacturing Inspections.

2

Components of the Laboratory Managerial and Administrative Systems Sub Element (MS)

Description of the Laboratory Managerial and Administrative Systems Sub Element

The Laboratory Managerial and Administrative Systems are those sub elements that provide the infrastructure for efficient and compliant operations of an analytical laboratory. It is sometimes difficult to directly relate items within this sub element to the Current Good Manufacturing Practices (CGMPs). However, deficiencies within this sub element, which cannot be directly attributed to the CGMPs, can lead to compliance failures. For example, 21 CFR Sections 210 and 211 do not specifically require that someone be assigned the duties of training manager. However, failing to have an individual responsible for training and consequently not having an effective training program can degrade the overall quality of data that is generated by laboratory personnel, and US Food and Drug Administration (US FDA) has cited companies for not having a training program for laboratory personnel [1].

The Laboratory Managerial and Administrative Systems sub element includes at least six individual topics. A laboratory that is in compliance with CGMPs should at least address each of these topics (as applicable). These topics include (i) Organizational Structure and Roles and Responsibilities, (ii) Training and Qualification, (iii) Laboratory Budgeting, Purchasing, and Requisition, (iv) Laboratory Administration and Operations, (v) Laboratory Chemicals, Solutions, Reagents and Supplies, and (vi) Laboratory Reference Standards and Solutions.

These six topics are listed within the text of the chapter along with some suggestions on what items should be addressed in each topic. The list of items under each topic is fairly comprehensive but may not be totally inclusive of all those components, which may constitute the sub element.

Laboratory Control System Operations in a GMP Environment, First Edition. David M. Bliesner.

Companion website: www.wiley.com/go/Bliesner/LabControl_GMPEnvironment

Contents of the Sub Element

The six topics mentioned earlier are presented in Table 2.1 in the form of questions related to each topic. For each topic, additional details, or "points to consider," are offered along with notes and clarifications as appropriate. The questions, notes, and clarifications are all based on actual inspectional findings from Regulatory Inspectors or points that have been found to directly or indirectly impact compliant and efficient real-world Laboratory Control System operations.

It is envisioned that the below table may be used as a basis for:

- Evaluating your own laboratory's managerial and administrative systems
- Upgrading your own laboratory's managerial and administrative systems
- Establishing your own laboratory's managerial and administrative systems if it is a new laboratory or if you are making substantial changes to an existing Laboratory Control System

Table 2.1 Components of the laboratory managerial and administrative systems sub element.

#	Laboratory managerial and administrative systems sub element topic	Notes and clarifications
2.1	***Organizational Structure and Roles and Responsibilities***	
2.1.1	Are current organization charts available for review and are they accurate?	Organizational charts will be one of the first things an auditor will request when your laboratory is being inspected. Make sure the document is current and correct in order to make a good first impression.
2.1.2	Is there a policy/procedure that defines the responsibility and authority of the Quality Control (QC) unit?	
2.1.3	Have the responsibilities of each functional group and supervisor been clearly defined, including testing and operational requirements, Standard Operating Procedures (SOPs), and all other critical functions?	

(Continued)

Table 2.1 (Continued)

#	Laboratory managerial and administrative systems sub element topic	Notes and clarifications
2.1.4	Is an organizational structure in place which is properly staffed to assure that all required testing/monitoring and support activities are performed?	Section 211.25 Personnel qualifications of the Good Manufacturing Practices regulations states "(c) There shall be an adequate number of qualified personnel to perform and supervise the manufacture, processing, packing, or holding of each drug product." This applies to the laboratory and failing to meet this requirement is often the root cause of many issues which occur in poorly operated QC laboratories.
2.1.5	Is the span of control and authority assigned to the QC laboratory adequate to allow proper execution of these activities?	
2.1.6	Is the ratio of supervisors to analyst appropriate? (A ratio of no more than 1 : 8 is recommended)	If a supervisor has to supervise more than eight employees directly, the effectiveness of their ability to do so decreases dramatically.
2.1.7	Are the roles and responsibilities for each position within the laboratory clearly defined?	
2.1.8	Are the roles and responsibilities for each position in the organizational chart defined in formal, controlled, accurate, and current job descriptions?	Up until recently, job descriptions were considered the purview of the human resources department and therefore did not fall under the umbrella of GMPs. This has now changed, and the general expectation of regulatory agencies is that job descriptions should be maintained and controlled like any other GMP document.
2.1.9	Are signature authority, responsibility, and accountabilities appropriate and clearly defined?	
2.1.10	Do systems exist to enhance communications, understanding, and working relationships between laboratory and Quality Assurance (QA) personnel?	A healthy, respectful relationship between QC and QA will do wonders for everyone's ability to manufacture, test, and, ship product. QC managers should encourage QA personnel to spend time in the laboratory and teach QA personnel as much about product testing and analytical techniques as they can.

(Continued)

Table 2.1 (Continued)

#	Laboratory managerial and administrative systems sub element topic	Notes and clarifications
2.1.11	Does a personnel performance evaluation system exist, which tracks laboratory personnel strengths and weaknesses and establishes a means to improve technical, administrative, and managerial skills?	Performance evaluation should not be a one-time-a-year event. A good leader will continually evaluate and assist their employees to become the best that they can be.
2.1.12	Does a master testing schedule or similar document(s) exist to ensure smooth and efficient work flow, and minimize laboratory personnel over-commitment?	
2.1.13	Are current copies of CV's and resumes for all personnel, including consultants, available for review?	
2.2	***Training and Qualification***	
2.2.1	Have the educational, training, and work experience requirements for each laboratory position been clearly defined and do they reflect current standards in the industry?	
2.2.2	Is the requirement for training and qualification clearly described in an SOP or similar document for all laboratory managers, supervisors, analysts, support, temporary staff, and consultants?	Training is defined as an activity designed to provide the skills and/or knowledge that individuals need to perform assigned job-related duties and responsibilities. Qualification is defined as the process, including documentation and approval, whereby a person is trained and demonstrates proof-of-proficiency in specified tests, tasks, procedures, or techniques.
2.2.3	Has a training and qualification curriculum been developed for each position that clearly identifies training requirements for all required SOPs and Policies, Test Methods, Safety Procedures, and the GMPs as well as all other internal and external courses or programs?	

(Continued)

Table 2.1 (Continued)

#	Laboratory managerial and administrative systems sub element topic	Notes and clarifications
2.2.4	Is there a system in place (e.g. electronic, paper-based, or hybrid) to capture, track, and manage training and qualification of laboratory personnel?	Electronic training systems should have a set of standard file + curriculum for each unique position on the organizational chart. For example, the position for a laboratory technician responsible for glassware washing should be loaded into the training system as if it was a real person. It will include all the necessary SOPs required for the position, qualification tasks, safety training, etc. This "standard position" will serve as a template. It then lays the basis for an individual's training record, which will grow and change as the employee is trained and qualified in that position.
2.2.5	Does each employee have a training file or similar record within the training system?	
2.2.6	Are the training histories for each individual employee kept current?	
2.2.7	Have all laboratory personnel been properly trained and qualified?	This point seems obvious; however during auditing it is not unusual to see someone who was "thrown into a position" because of labor shortages, who has never been trained on the tasks they are preforming in the laboratory.
2.2.8	Are the training histories for each individual employee readily available for review?	
2.2.9	Is SOP training conducted in a fashion other than "read and understand" as appropriate?	Unfortunately, organizations have a tendency to perform all SOP training in this fashion. It is not effective in many circumstances and the practice should be restricted to review of minor changes on previously trained-on procedures.

(Continued)

Table 2.1 (Continued)

#	Laboratory managerial and administrative systems sub element topic	Notes and clarifications
2.2.10	Are metrics, related to the execution of training, collected, analyzed, and used to determine the effectiveness of the employee training system in general?	
2.2.11	Do the metrics used to evaluate training system effectiveness include the following:	Metrics can take many different forms; these are just a few suggestions.
2.2.11.1	Number of courses completed versus courses scheduled?	
2.2.11.2	Percent of training that is overdue?	
2.2.11.3	Evaluation of training course content by attendees?	
2.2.11.4	Evaluation of instructor effectiveness by course attendees?	
2.2.11.5	Evaluation of test questions to determine their appropriateness and effectiveness in evaluating knowledge?	
2.2.12	Is there an analyst qualification program (e.g. proof-of-proficiency for laboratory skills) in place?	
2.2.13	Has an individual been designated as the training coordinator or manager?	If the training manager or coordinator is not someone with a laboratory background, make sure they are supported by someone who has a science background with hands-on laboratory experience.
2.2.14	Is there evidence of management support for training and training programs?	
2.2.15	Does a formal training budget exist?	
2.2.16	Do vendor and consultant training records exist?	

(Continued)

Table 2.1 (Continued)

#	Laboratory managerial and administrative systems sub element topic	Notes and clarifications
2.2.17	Is there evidence that newly hired employees are evaluated for skill level, competency, and proof-of-proficiency with respect to laboratory, supervisory, and managerial skills as appropriate?	One of the adages with respect to finding qualified personnel is we hire too quickly and fire too slowly. Take your time when hiring and verify, to the best of your organization's ability that who you are hiring truly does possess the education, training, and experience to do the job you need them to do. Also be aware that no matter how competent and experience a new hire is, it will most likely take them at least six months to come fully up-to-speed within your organization. And if you doubt this, you are fooling yourself.
2.2.18	Is there evidence that newly hired employees are evaluated for their language and literacy skills?	In culturally diverse workplaces, it is not unusual to encounter personnel who are not native speakers, readers, or writers in the primary language used at the facility. Since it is a GMP requirement that procedures exist and such procedures shall be followed, personnel should be evaluated for their ability to read, understand, and execute the SOPs for which are required to be followed in performance of their job responsibilities.
2.2.19	Is there a formal training schedule in place and is it being executed?	
2.2.20	Is there evidence of employee re-training and re-qualification?	The GMPs in Section 211.25 (a) state "Training in current good manufacturing practice shall be conducted by qualified individuals on a continuing basis and with sufficient frequency to assure that employees remain familiar with CGMP requirements applicable to them." Although this section of the GMPs is targeted to GMP training, it should be expanded to cover the broader arena of job skills in general.

(Continued)

Table 2.1 (Continued)

#	Laboratory managerial and administrative systems sub element topic	Notes and clarifications
2.2.21	Do managers, supervisors, or training coordinators/managers periodically meet with each employee and review the employee's training record and discuss training and qualification effectiveness and needs?	
2.2.22	Is there a positive attitude with respect to training on the part of management and laboratory personnel?	Experience has shown that most employees will openly discuss their pleasure (or displeasure) in the content, effectiveness, and frequency of training and qualification. Employees are also the best to know what needs to be done to upgrade existing training efforts or what new efforts need to be undertaken.
2.3	***Laboratory Budgeting, Purchasing, and Requisition***	
2.3.1	Are levels of approval defined and appropriate for the purchasing and requisitioning needs of the laboratory?	There should be more than one person who can sign, initiate requisitions, and sign purchase orders in the laboratory. This needs to be defined in writing.
2.3.2	Are items purchased from qualified vendors (e.g. reagents, standards, instruments, etc.)?	
2.3.3	Does the laboratory have a defined yearly expendables budget?	
2.3.4	Is the expendables budget derived from existing data and takes into consideration projected changes in manufacturing volume and its impact on testing workloads?	
2.3.5	Does the laboratory have a clearly defined and functional capital expenditure budgeting process?	Capital expenditures include purchasing pieces of equipment and instrumentation such as HPLCs, GCs, balances, etc.
2.3.6	Is the capital budget derived from existing data and take into consideration new product introductions, changes in manufacturing workloads, etc.?	The age of equipment should also be a consideration. Equipment and often the software used to run it need to be replaced from time-to-time. Aging equipment can lead to unnecessary laboratory investigations, which are time-consuming and often not value-added.

(Continued)

Table 2.1 (Continued)

#	Laboratory managerial and administrative systems sub element topic	Notes and clarifications
2.3.7	Is there a change control system in place to ensure that the purchasing department cannot "swap out" supplies or vendors without involving the laboratory purchasing personnel to prevent inappropriate substitution of materials or equipment?	For example, instances have been observed where the purchasing department, in an effort to save money, has switched vendors who make syringes, which are used to remove samples from dissolution baths. The result was extraneous peaks in the chromatography because the less expensive syringes included a lubricant in the syringe barrel, which was not present in the original validated method syringes.
2.3.8	Does the laboratory have a defined process for requesting and receiving approval to hire new personnel?	
2.3.9	Does the laboratory have specific personnel assigned to purchasing and budgeting activities?	
2.3.10	Do the personnel assigned to purchasing and budgeting have the proper skills and training to perform planning, budgeting, and purchasing activities?	
2.4	***Laboratory Administration and Operations***	
2.4.1	Does the laboratory possess work flow or process diagrams, at least for its primary operations, such as sample management, testing, data review and management, and reporting of results?	Work flow diagrams or process maps are invaluable and lay the basis for efficient operations for most all tasks performed in the laboratory. See the Chapter 2 Appendix (www.wiley.com/go/Bliesner/LabControl_GMPEnvironment) for an example work flow diagram-template.
2.4.2	Does the laboratory have a general administrator, office manager, or similar administrative support person?	
2.4.3	Does the administrator, office manager, or similar support person perform or supervise the following activities:	
2.4.3.1	Managing work orders for office and laboratory repairs?	

(Continued)

Table 2.1 (Continued)

#	Laboratory managerial and administrative systems sub element topic	Notes and clarifications
2.4.3.2	Overseeing stockroom operations?	
2.4.3.3	Supervising requisition of administrative and operational supplies?	
2.4.3.4	Managing external contractors and contracts?	
2.4.3.5	Managing/enhancing communication between the laboratory and internal/external customers?	
2.4.3.6	Supervising laboratory safety training and compliance?	
2.4.3.7	Monitoring overtime billing?	
2.4.3.8	Initiating capital equipment purchases?	
2.4.3.9	Attending production planning meetings?	
2.4.3.10	Representing the laboratory at Quality Council, Quality Review, or similar meetings where site-wide quality metrics are reviewed?	It is critical that the QC laboratory be engaged with the other Quality Systems within the organization, particularly Manufacturing and QA. It is not unusual for the QC laboratory to become isolated from the other systems, which leads to an underappreciation of the challenges that exist for running a laboratory. It can also lead to unrealistic expectations regarding the volume of testing, which can be achieved and testing turnaround times, especially at the higher levels of management. Personnel at these levels rarely have laboratory experience and need to be respectfully educated on the nuances and challenges of running a QC laboratory.

(Continued)

Table 2.1 (Continued)

#	Laboratory managerial and administrative systems sub element topic	Notes and clarifications
2.4.3.11	Collecting, analyzing, reporting, and presenting quality metrics related to the QC laboratory (e.g. laboratory errors, challenges related to test methods, out-of-specification (OOS) and out-of-trend investigations (OOT) investigations, equipment issues, etc.)?	See the Chapter 2 Appendix (www.wiley.com/go/Bliesner/LabControl_GMPEnvironment) for some example metrics.
2.4.3.12	Collecting, analyzing, reporting, and presenting operational metrics related to the QC laboratory (e.g. right-first-time on testing, testing turnaround times, testing backlogs, etc.)?	
2.4.3.13	Managing regulatory agency inspections?	
2.4.3.14	Managing internal and external audits?	
2.4.4	Does the administrator, office manager, or similar support person perform or supervise keeping current with FDA and regulatory related items such as:	FDA has several e-mails feeds to which you can subscribe, including notification of Warning Letters, Guidance Document updates, etc. through the Subscription Management Center: https://www.fda.gov/about-fda/contact-fda/get-email-updates You will receive automatic e-mail notifications to the topics with which you wish to stay current.
2.4.4.1	Recent Warning Letters?	
2.4.4.2	Recently posted Form 483 observations?	Electronic reading room for the Centers for Drug Evaluation and Research (CDER) will sometimes post Form 483s from inspections in the website section "Frequently requested or proactively posted compliance records" https://www.fda.gov/drugs/guidance-compliance-regulatory-information/cder-foia-electronic-reading-room

(Continued)

Table 2.1 (Continued)

#	Laboratory managerial and administrative systems sub element topic	Notes and clarifications
2.4.4.3	New FDA initiatives?	FDA has recently become very active and is spearheading numerous activities which need to be tracked and understood.
2.4.4.4	New or updated Guidance Documents?	https://www.fda.gov/regulatory-information/search-fda-guidance-documents
2.4.4.5	Keeping current with the USP/NF, BP, JP, EP, etc.?	Organizations are encouraged to purchase online subscriptions to the USP/NF and other pharmacopeia as appropriate.
2.4.4.6	Keeping current with industry groups such as American Chemical Society (ACS), American Association of Pharmaceutical Scientists (AAPS), Parenteral Drug Association (PDA), etc.?	
2.4.5	Does the administrator, office manager, or similar support person perform or supervise keeping current with industry trends by monitoring or subscribing to:	
2.4.5.1	The "Fierce" family of industry blogs (www.fiercepharma.com)?	The "Fierce Family" of blogs includes FiercePharma, FiercePharma Manufacturing, FiercePharmaMarketing, and others.
2.4.5.2	PIC/S? (Pharmaceutical Inspection Convention Pharmaceutical Inspection Co-Operation Scheme)	
2.4.5.3	WHO? (World Health Organization)	
2.4.5.4	ICH? (International Council for Harmonisation of Technical Requirements for Pharmaceuticals for Human Use)	https://www.ich.org/home.html
2.4.6	Does the administrator, office manager, or similar support person supervise administrative service personnel such as maintenance and calibration, laboratory computers, documentation section, training section, etc.?	

(Continued)

Table 2.1 (Continued)

#	Laboratory managerial and administrative systems sub element topic	Notes and clarifications
2.5	***Laboratory Chemicals, Solutions, Reagents, and Supplies***	
2.5.1	Is there a written procedure for receipt and storage of chemicals and reagents?	
2.5.2	Are chemicals and reagents properly labeled with date of receipt, date opened, and expiration and re-test dates?	
2.5.3	Are laboratory prepared reagents and solutions properly identified (e.g. chemical name or symbol, concentration, date of preparation, initials of the analyst who prepared it, and expiration date)?	Unlabeled solutions are an FDA form 483 favorite citation.
2.5.4	Are records available which document preparation and standardization of volumetric solutions?	
2.5.5	Is the frequency of standardization of various reagents described? Is it consistent with the current USP?	
2.5.6	Are standardized reagents properly stored to assure integrity?	
2.5.7	Does a procedure exist, which describes the generation/purchase, storage, and handling of reagent grade water for laboratory use?	Laboratory water, when opened (exposed to air) will start to grow microbes almost immediately. Therefore, experience has shown that lab water should not be held for greater that 24 hours. If the water is used to prepare chromatographic samples, extraneous peaks resulting from the microbial contamination are possible. This means that water stored in squirt bottles needs to be changed daily.
2.5.8	Are there procedures that describe the preparation of laboratory reagents and cultures?	
2.5.9	Are there procedures that describe the maintenance of laboratory reagents and cultures?	

(Continued)

Table 2.1 (Continued)

#	Laboratory managerial and administrative systems sub element topic	Notes and clarifications
2.6	***Laboratory Reference Standards and Solutions***	
2.6.1	Is there a written procedure for ordering and receipt of compendial reference standards and non-compendial reference standards?	
2.6.2	Are the primary standards of the current lot listed in the USP, EP, or JP?	
2.6.3	Is the receipt of the standards logged?	
2.6.4	Are all standards labeled with name, source, lot number, and expiration date?	
2.6.5	Do written procedures include provisions for prevention of contamination of primary standards?	
2.6.6	Are standards stored in a secured area under environmentally controlled and monitored conditions?	
2.6.7	Are procedures for assuring standard integrity available?	
2.6.8	Are working or "house" standards checked against primary standards at appropriate intervals?	
2.6.9	Are stock solutions appropriately identified, and is their expiration based on their true stability?	
2.6.10	Do procedures exist for the certification and use of non-commercially available reference standards?	
2.6.11	Do certificates of analysis exist for all reference standards and are these certificates stored as controlled documents?	
2.6.12	Have provisions been made for handling, controlled substance reference standards?	This refers to Drug Enforcement Agency (DEA) scheduled materials (e.g. methamphetamine, methadone, etc.).
2.6.13	Does the Reference Standard SOP address (map) proper handling of controlled substance reference standards?	

Tools and Templates

The following are provided in electronic format as Example Tools and Templates, which can be found in the Chapter 2 Appendix (www.wiley.com/go/Bliesner/LabControl_GMPEnvironment):

- Checklist for Evaluating the Laboratory Managerial and Administrative Systems Sub Element
- Example-Template Job Description
- Example-Template Work Flow Diagram
- An Example Listing of Quality Compliance and Operational Laboratory Metrics

Reference

1 US FDA (2013). Warning Letter WL: 320-13-12, 25 March 2013.

3

Components of the Laboratory Documentation Practices and Standard Operating Procedures Sub Element (OP)

Description of the Laboratory Documentation Practices and Standard Operating Procedures Sub Element

Laboratory Documentation Practices and Standard Operating Procedures form a critical component of the overall Laboratory Control System. Much of the essence of Current Good Manufacturing Practices (CGMPs) involves the generation, review, approval, revision, and control of procedures and data. In compliance parlance, "If it isn't written down, it didn't happen."

The Laboratory Documentation Practices and Standard Operating Procedures sub element mirrors the Laboratory Control System itself. Therefore, there are 10 individual parts included in the sub element. A laboratory, which is in compliance with CGMPs, should have the procedures and their corresponding outputs addressed in each of these areas (as applicable). These parts include (i) Standard Operating Procedures (SOPs) Related to the Laboratory Managerial and Administrative Controls Sub Element, (ii) SOPs Related to the Laboratory Documentation Practices and Standard Operating Procedures Sub Element, (iii) SOPs Related to the Laboratory Equipment Sub Element, (iv) SOPs Related to the Laboratory Facilities Sub Element, (v) SOPs Related to the Method Validation and Method Transfer Sub Element, (vi) SOPs Related to the Laboratory Computers Sub Element, (vii) SOPs Related to the Laboratory Investigations Sub Element, (viii) SOPs Related to Data Governance and Data Integrity Sub Element, (ix) SOPs Related to the Stability Program Sub Element, and (x) SOPs Related to the General Laboratory Compliance Practices Sub Element.

The procedures and sub sections described in the succeeding text are fairly comprehensive in the number and type. However, your organization may require additional procedures to round out your Laboratory Documentation Practices and Standard Operating Procedures sub element. Additionally, you will have to provide the appropriate sub sections and detail to your own procedures when they are written or modified.

Laboratory Control System Operations in a GMP Environment, First Edition. David M. Bliesner.

Companion website: www.wiley.com/go/Bliesner/LabControl_GMPEnvironment

Contents of the Sub Element

Table 3.1 is a model master laboratory SOP index organized by laboratory control system sub element. It lists 63 SOP titles that may be more or less what you would find in a typical quality control (QC) laboratory.

Table 3.2 is an expanded version of the SOP index but includes additional detail, in the form of sub section titles, or notes and clarifications, as appropriate.

It is envisioned that the below tables may be used as a basis for:

- Evaluating your own laboratory's documentation practices and standard operating procedures
- Upgrading your own laboratory's documentation practices and standard operating procedures
- Establishing your own laboratory's documentation practices and standard operating procedures if it is a new laboratory or if you are making substantial changes to and existing Laboratory Control System

Table 3.1 Model master laboratory SOP index.

#	1.0-MS	SOPs Related to the Laboratory Managerial and Administrative Controls Sub Element
1	SOP-MS-1.1	*Organizational Structure, Roles, Responsibilities and Performance Evaluation*
2	SOP-MS-1.2	*Work Flow, Sample Management, Tracking, Trending, and Release of Analytical Data*
3	SOP-MS-1.3	*Laboratory Training and Qualification Program*
4	SOP-MS-1.4	*Laboratory Safety Program*
5	SOP-MS-1.5	*Laboratory Chemicals, Solutions, Reagents and Supplies*
6	SOP-MS-1.6	*Reference Standards, In-House Standards, and Standard Solutions*
7	SOP-MS-1.7	*Laboratory Administration and Operations*
8	SOP-MS-1.8	*Special Testing and Procedures*
	2.0-OP	**SOPs Related to the Laboratory Documentation Practices and Standard Operating Procedures Sub Element**
9	SOP-OP-2.1	*Laboratory Document Control System*
10	SOP-OP-2.2	*Laboratory Documentation Practices*
11	SOP-OP-2.3	*Laboratory Standard Operating Procedures, Test Methods and Specifications*

(Continued)

Table 3.1 (Continued)

#	1.0-MS	SOPs Related to the Laboratory Managerial and Administrative Controls Sub Element
	3.0-LE	**SOPs Related to the Laboratory Equipment Sub Element**
12	SOP-LE-3.1	*Overview of the Laboratory Equipment Program*
13	SOP-LE-3.2	*Assessing Laboratory Equipment Needs*
14	SOP-LE-3.3	*Sourcing and Purchasing Laboratory Equipment*
15	SOP-LE-3.4	*Laboratory Equipment Inventory, Inventory Control and Usage Tracking*
16	SOP-LE-3.5	*Installation, Operational and Performance Qualification (IQ/OQ/PQ) of Laboratory Equipment*
17	SOP-LE-3.6	*Equipment Maintenance and Calibration*
18	SOP-LE-3.7	*Repairing Laboratory Equipment*
19	SOP-LE-3.8	*Laboratory Equipment Lifecycle Management*
20	SOP-LE-3.9	*United States Pharmacopeia (USP) Instrument Groupings and Equipment Related Guidance Chapters*
21	SOP-LE-3.10	*Laboratory Glassware*
	4.0-LF	**SOPs Related to the Laboratory Facilities Sub Element**
22	SOP-LF-4.1	*Laboratory Building and Facility Maintenance*
23	SOP-LF-4.2	*Laboratory Water Systems*
24	SOP-LF-4.3	*Laboratory Building and Facilities Security and Access Control*
25	SOP-LF-4.4	*Temperature and Humidity Monitoring Systems and Chart Recorders*
26	SOP-LF-4.5	*Fume Hood Use, Cleaning, and Maintenance*
27	SOP-LF-4.6	*Disaster Recovery Plan*
	5.0-MV	**SOPs Related to the Methods Validation and Method Transfer Sub Element**
28	SOP-MV-5.1	*Analytical Test Method Development*
29	SOP-MV-5.2	*Analytical Method Validation*
30	SOP-MV-5.3	*Analytical Test Method Transfer*
31	SOP-MV-5.4	*Verification of Compendial Procedures*
	6.0-LC	**SOPs Related to the Laboratory Computers Sub Element**
32	SOP-LC-6.1	*Electronic Records and Signatures*
33	SOP-LC-6.2	*Electronic Records Storage, Backup, Archival, and Restoration*
34	SOP-LC-6.3	*General Procedures for Computer System Validation*
35	SOP-LC-6.4	*Chromatographic Data Acquisition Software*
36	SOP-LC-6.5	*Electronic Laboratory Notebooks (ELN)*
37	SOP-LC-6.6	*Laboratory Information Management System (LIMS)*
38	SOP-LC-6.7	*Computer System Change Control Procedures*
39	SOP-LC-6.8	*Validation of Spreadsheets*
40	SOP-LC-6.9	*Validation of Databases*

(Continued)

Table 3.1 (Continued)

#	1.0-MS	SOPs Related to the Laboratory Managerial and Administrative Controls Sub Element
41	SOP-LC-6.10	*Power Failure Recovery Procedures for Computers*
42	SOP-LC-6.11	*Disaster Recovery of Electronic Data and Computer Equipment*
43	SOP-LC-6.12	*Computer System Security*
44	SOP-LC-6.13	*Operational Maintenance of Computer Systems and Software*
	7.0-LI	**SOPs Related to the Laboratory Investigations Sub Element**
45	SOP-LI-7.1	*Conducting, Documenting and Reporting Laboratory Investigations (LIRs), Out of Specification (OOS) and Out of Trend Investigations (OOT)*
46	SOP-LI-7.2	*Procedures for Addressing Unknown or Extraneous Chromatographic Peaks*
47	SOP-LI-7.3	*System Suitability Failures: Capture, Tracking, and Patterning*
	8.0 DI	**SOPs Related to Data Governance and Data Integrity Sub Element**
48	SOP-DI-8.1	*Overview of Data Governance and Data Integrity*
49	SOP-DI-8.2	*Data Sources and Data Mapping*
50	SOP-DI-8.3	*Application of Hazard Analysis Critical Control Points (HACCP) to Laboratory Data Integrity*
51	SOP-DI-8.4	*Personnel Compliance Program for Insuring Laboratory Data Integrity*
	9.0-SB	**SOPs Related to the Stability Program Sub Element**
52	SOP-SB-9.1	*An Overview of the Stability Program*
53	SOP-SB-9.2	*Initiating Stability Studies (Study Requests and Protocol Generation)*
54	SOP-SB-9.3	*Activating Stability Studies (Sample Set-Down)*
55	SOP-SB-9.4	*Time Zero Testing, Review, and Reporting*
56	SOP-SB-9.5	*Withdrawal of Drug Product from Stability Chambers (Sample Pull)*
57	SOP-SB-9.6	*Stability Sample Testing and Results Reporting*
58	SOP-SB-9.7	*Stability Data Analysis, Reporting, and Shelf-Life Assignment*
59	SOP-SB-9.8	*Administration of the Stability Program and Stability Studies*
60	SOP-SB-9.9	*Additional Types of Stability Studies: Shipping Studies, Thermal Excursion Studies, Bulk and Blend Hold Time Studies, Post Approval Change Studies and Reserve Samples*
61	SOP-SB-9.10	*Selection, Qualification, Calibration and Maintenance of Stability Chambers*
	10.0-CP	***SOPs Related to the General Compliance Practices Sub Element***
62	SOP-CP-10.1	*Laboratory Housekeeping and General Appearance*
63	SOP-CP-10.2	*Periodic Laboratory-Led CGMP Compliance Inspections*

Table 3.2 Model master laboratory SOP index with details.

SOP #	Section #	Title or header	Notes and clarifications
1.0-MS		**SOPs Related to the Laboratory Managerial and Administrative Controls Sub Element**	See Chapter 2: Components of the Laboratory Managerial and Administrative System Sub Element for additional detail.
SOP-MS-1.1		*Organizational Structure, Roles, Responsibilities, and Performance Evaluation*	Many QC laboratories do not spell out in writing the topics listed in this procedure. They instead are handled as common knowledge which often require someone "who knows them" to explain how they are executed in the laboratory.
	Section 1.1.1	Organizational Structure, Responsibility and Authority of the Quality Control Laboratory	
	Section 1.1.2	The Role of Quality Assurance and QA Oversight of Laboratory Operations	This topic is rarely addressed in writing in the QC laboratory. However, it is critical that QA and QC have a professional, respectful and productive relationship. Laboratories often get into trouble if QA is not actively involved in compliance oversight in the QC laboratory. QA should be finding (and helping fix) laboratory compliance challenges so that when Regulators show up for inspections, there will be minimal complications. QA has the ultimate authority and responsibility for oversight of the QC laboratory operations to include review and approve all results before product is released to market and review and approval of laboratory procedure. This fact needs to be spelled out in this procedure.
	Section 1.1.3	Position Descriptions and Job Expectations for Personnel in the Quality Control Laboratory	Once the realm of the Human Resource Departments, Job Descriptions are now a regulatory expectation.

(Continued)

Table 3.2 (Continued)

SOP #	Section #	Title or header	Notes and clarifications
	Section 1.1.4	Signatures and Signature Authority	It should be documented in writing what documents require approval signatures and who in the laboratory has the authority to sign them. This ranges from administrative tasks such as purchase orders to compliance tasks such as approval of protocols. There should be a list of all personnel in the laboratory and examples of their full signatures and initials. The list can also include their job titles and job responsibilities.
	Section 1.1.5	Terms and Acronyms Used in the Quality Control Laboratory	
	Section 1.1.6	Performance Evaluation and Goal Setting for Quality Control Laboratory Personnel	
	Section 1.1.7	Laboratory Leadership Succession Plan	Succession in this case means who takes the duties of a lab director, manager, supervisor, etc. in their absence. This succession plan should address planned absences such as vacations and offsite training, to unplanned absences such as illness or family emergencies.
SOP-MS-1.2		*Work Flow, Sample Management, Tracking, Trending, and Release of Analytical Data*	Work flow maps or process diagrams lay the basis for efficient and compliant laboratory operations.
	Section 1.2.1	Master Testing Schedule Creation, Upkeep, and Work Flow Management	
	Section 1.2.2	Creation and Execution of Sampling Plans	Depending upon the organization sampling of samples from manufacturing may be performed by QA, QC or manufacturing personnel. Regardless of who performs the sampling, the SOP should provide instructions so that the selected samples are representative of the entire lots(s) and that no "cherry picking" has resulted.

	Section 1.2.3	Sample Labeling, Receipt, Log-In, Storage, Issuance, Tracking, Retention, and Disposal Procedures	Sample logs may either be electronic or manual. Samples should be secured in a controlled and limited access area until they are disposed.
	Section 1.2.4	Data Flow, Control, Review, Approval, Retention and Archival in the Quality Control Laboratory	The subject of limiting access to stored data should be addressed here, either by direct discussion, or reference to a separate procedure associated with the Data Governance and Data Integrity sub element. The same applies to the physical security of data (electronic or otherwise) to prevent damage from fire, water or other environmental hazards. The length of time that data and documents need to be retained should be defined in this procedure and should reflect current industry practice.
	Section 1.2.5	Project Management and Work Flow Tools: Systems and Software	
	Section 1.2.6	Tracking, Trending and Reporting Testing Results	This procedure should include directions for identification and reporting of new impurities and/or higher levels of previously known impurities.
	Section 1.2.7	Quality Control Laboratory Support for Annual Product Reviews	Extracting data from LIMS or another electronic data storage system is the most practical approach.
SOP-MS-1.3		*Laboratory Training and Qualification Program*	Training is defined as an activity designed to provide the skills and/or knowledge that individuals need in order to perform assigned job-related duties and responsibilities. Qualification is defined as the process, including documentation and approval, whereby a person is trained and demonstrates proof-of-proficiency in specified tests, tasks, procedures, or techniques.
	Section 1.3.1	Overview and Administration of the Laboratory Training and Qualification Program	
	Section 1.3.2	Creating and Managing the Training Budget	
	Section 1.3.3	Responsibilities of the Laboratory Training Manager	

(Continued)

Table 3.2 (Continued)

SOP #	Section #	Title or header	Notes and clarifications
	Section 1.3.4	GMP Compliance Training	Traditionally a "Once a Year" evolution, to comply with the GMPs, GMP compliance training should be an on-going evolution and taught periodically throughout the year. A good way to do this is periodically present FDA posted Form 483s and Warning Letters so lab personnel can see how other organizations have gotten into compliance troubles.
	Section 1.3.5	New Employee Training	
	Section 1.3.6	Standard Operating Procedure (SOP) Training	Should be more than "Read and Understand" which is the least effective means to train people on new or significantly modified SOPs.
	Section 1.3.7	Continuous Improvement, Refresher Training and Re-Qualification of Laboratory Personnel	
	Section 1.3.8	Creating Training and Qualification Curricula and Materials for Laboratory Training and Qualification	Format for training and qualification curricula should be defined and used by all personnel who teach or help qualify lab personnel.
	Section 1.3.9	Vendor and Consultant Training Requirements	All vendors who perform on-site services related to GMP activities should have official training records. The same is true for consultants.
SOP-MS-1.4		*Laboratory Safety Program*	Laboratory safety is not a GMP requirement. However, if there are safety issues in a laboratory, it is almost a given that there are GMP deficiencies as well. If you don't care about safety, you will not care about compliance.
	Section 1.4.1	General Laboratory Safety Program	The most important job for a laboratory supervisor is to make sure his or her people go home at the end of the day with the same body parts, in the same condition, as when they came to work that day.
	Section 1.4.2	Laboratory Safety Inspection Guidelines	

	Section 1.4.3	Chemical Hygiene Plan	A Chemical Hygiene Plan is a written program developed and implemented by the organization which sets forth procedures, equipment, personal protective equipment and work practices that are capable of protecting employees from the health hazards presented by hazardous chemicals used in that particular workplace.
	Section 1.4.4	Handling and Disposal of Solvent and Chemical Waste	Waste disposal must comply with local, State, and Federal standards.
	Section 1.4.5	Handling and Disposal of Non-Liquid Laboratory Waste	This includes broken glass, used HPLC and GC vials, disposable gloves, etc.
	Section 1.4.6	Recording and Reporting Injuries	
	Section 1.4.7	Testing of Eyewash and Emergency Showers	Eyewash stations need to be periodically flushed to ensure that there is no microbial or parasitic growth in the lines.
SOP-MS-1.5		*Laboratory Chemicals, Solutions, Reagents, and Supplies*	
	Section 1.5.1	Ordering, Purchasing, Receiving, Stocking, and Expiration Dating of Laboratory Reagents, Chemicals and Supplies	
	Section 1.5.2	Ordering, Purchasing, Receiving, Storage, and Replacement of Laboratory Glassware	QC laboratories are notorious for keeping old or worn glassware which needs to be replaced.
	Section 1.5.3	Cleaning and Care of Laboratory Glassware	Cleaning and maintaining glassware are perhaps the most important support functions in any laboratory. Unclean or improperly cleaned glassware can, and frequently does, lead to extraneous peaks in chromatograms which require investigations.
	Section 1.5.4	Inventory Control for Laboratory Reagents, Chemicals, and Supplies	
	Section 1.5.5	Preparation, Expiration Dating and Record Keeping for Prepared Reagents, Solutions, and Mobile Phases	
	Section 1.5.6	HPLC and GC Column Ordering, Receipt, Use, Storage, Care and Disposal	HPLC and GC columns should be stored in a secure fashion. A record of the number of injections and types of analyses they were used for should also be logged.

(Continued)

Table 3.2 (Continued)

SOP #	Section #	Title or header	Notes and clarifications
SOP-MS-1.6		*Reference Standards, In-House Standards, and Standard Solutions*	
	Section 1.6.1	Guidelines for Use of Primary Reference Standards and In-House Reference Standards in the Quality Control Laboratory	
	Section 1.6.2	Receipt, Review, and Use of Compendial Reference Materials and Other Reference Standards	
	Section 1.6.3	Qualification and Re-Qualification of Secondary and In-House Reference Standards	
	Section 1.6.4	Operation and Maintenance of the Reference Standard Library	
SOP-MS-1.7		*Laboratory Administration and Operations*	These duties are usually shared amongst several individuals. However, if at all possible, having a full-time laboratory administrative support person is invaluable.
	Section 1.7.1	General Laboratory Administrative Support	
	Section 1.7.2	Laboratory Budgets, Budgeting, and Capital Equipment Purchases	
	Section 1.7.3	Managing Regulatory Agency Inspections	There should be a standard plan and it should be periodically rehearsed.
	Section 1.7.4	Managing External (non-regulatory) Audits	
	Section 1.7.5	Managing Internal and Corporate Audits	

	Section 1.7.6	Capturing, Tracking, Reporting, and Analyzing Key Performance Indicators for the QC Laboratory	Includes both quality compliance related metrics and operational related metrics. A key metric is the number of samples awaiting testing or on backlog. If backlog is too great quality of test results may be challenged as well as product release.
	Section 1.7.7	Managing External Contractors and Contracts	
	Section 1.7.8	Initiating, Managing, and Closing Work Orders for Laboratory and Office Repairs	
SOP-MS-1.8		*Special Testing and Procedures*	
	Section 1.8.1	Receipt, Testing and Results Reporting for Customer Complaint Samples	
	Section 1.8.2	Ordering, Receiving, Storing, Issuing, Testing, and Chain of Custody of Controlled Drugs, Drug Substances and Standards	As applicable to your laboratory.
	Section 1.8.3	In-Process Samples and In-Process Testing	In-Process Testing may be performed by the QC laboratory, Manufacturing, or split/shared by both. If more than In-Process testing is more than simple tests (e.g. pH measurement, weight variation, hardness, thickness, etc.), then the In-Process methods must be validated (e.g. Blend uniformity, tablet core assay for coated tablets, etc.) Sampling plans must exist for obtaining In-Process samples. In-Process test results must have pre-determined acceptance criteria and be documented in the In-Process test specification.

(Continued)

Table 3.2 (Continued)

SOP #	Section #	Title or header	Notes and clarifications
2.0-OP		**SOPs Related to the Laboratory Documentation Practices and Standard Operating Procedures Sub Element**	
SOP-OP-2.1		*Laboratory Document Control System*	The Laboratory Document Control sub element is an extension of the corporate and site-wide document control system which falls under the Quality System of the FDA six system model for quality.
	Section 2.1.1	Overview of the Laboratory Document Control System	This is sometimes referred to as the SOP for SOPs. It should address the hierarchy of the SOP system for the company including Level I and Level II (e.g. Corporate and Site guidance documents and policies). Laboratory Documentation includes all records generated or used to support testing and to comply with the GMPs. This includes, at least: test methods, operational SOPs, specifications, calibration, and maintenance procedures, forms, electronic raw data and final reported results (hard copies and electronic), lab notebooks, worksheets, logbooks, instrument printouts, reports, investigation and deviation reports and supporting data, CAPAs, etc.
	Section 2.1.2	Use of Standard Operating Procedures	This SOP clarifies points regarding use of SOPs and STMs (Standard Test Methods) during sample testing in the lab. The procedure should also address the fact that all procedures be current, clearly written and accessible to all laboratory personnel, as appropriate. The method used for testing, including revision number, needs to be documented in the notebook, worksheet or LIMS.
	Section 2.1.3	Initiation, Creation, Review, Approval, Control, and Maintenance of Laboratory Documentation	This SOP addresses laboratory document change control. The procedure should also discuss periodic review of all SOPs to ensure that they are consistent with current company and industry practices. The review occurs at least every two years (biennial).

	Section 2.1.4	Maintaining the Master Laboratory Control System SOP and Specification Index	The index must be continually updated to reflect changes, additions and deletions and accurately reflect procedures and specifications in current use.
	Section 2.1.5	Issue and Use of Single Sheet or Non-Bound Notebooks	
	Section 2.1.6	Issue and Use of Hardbound Notebooks	
	Section 2.1.7	Issue, Use, and Control of Laboratory Forms	If forms are available electronically, controls must be in place which prevents unrestricted printing of the form.
SOP-OP-2.2		*Laboratory Documentation Practices*	
	Section 2.2.1	Policy for Contemporaneous and Accurate Documentation in the QC Laboratory: Good Documentation Practices (GDP)	This section must align with Site-wide GDP procedures and address the basics such as prohibition of the use of scratch paper, proper error correction, use of permanent ink, requirements for signing and dating entries, requirement for contemporaneous and complete entries and second person review and approval including a signature. The second person should be trained and qualified on the testing and data review requirements in order to serve as a reviewer. All testing records should be clear and completed and be readily retrievable.
	Section 2.2.2	Write-Up and Reporting of Analytical Test Results	SOP should include requirements for including or referencing all outputs such as chromatograms, spectra, etc. Example calculations should be included in the write-ups, which include units and explanation of corrections factors as appropriate. A statement of compliance with the specification (e.g. numeric, pass, describing result, etc.) should be documented. If entries are made or transcribed into LIMS they must be verified if performed manually, with all entries in the LIMS system being approved and evidence of approval being captured. Also, these results need to be verified against the established specifications. Quality Assurance must be involved in reviewing and approving final results to determine if they meet specifications.

(Continued)

Table 3.2 (Continued)

SOP #	Section #	Title or header	Notes and clarifications
	Section 2.2.3	Guidelines for HPLC Set-Up, Sample, and Standard Queues and Calculations	Sample queues refer to the order in which system suitability standards, samples and calibration standards are organized for a chromatographic run. Sample queues should be codified in a procedure (or the individual method) and be consistent for the various types of analyses such as assay, impurities, identification, etc. This section should address or reference a separate SOP regarding guidelines for establishing Chromatographic Data System (CDS) integration parameters, restrictions on reintegration and manual integration.
	Section 2.2.4	Reviewing Chromatographic Data	
	Section 2.2.5	Corrective and Preventive Action (CAPA) Procedures	
	Section 2.2.6	Conventions for Reporting Analytical Data: Raw Data, Rounding, Significant Figures and Statistical Treatment of Results	The definition of raw data should be defined in this section, or refer to the definition in a separate procedure such as one associated with the Data Integrity sub element.
SOP-OP-2.3		*Laboratory Standard Operating Procedures, Test Methods, and Specifications*	All of these document types must be reviewed and approved by QA.
	Section 2.3.1	Creation, Use, Control and Maintenance of Analytical Test Methods	All test methods, for API, finished product, raw material, etc. must be validated or verified, and the most current version used during testing. This list can be substantial depending upon the number and types of tests required and the number of products being manufactured and may include: Description, Assay, Impurities/Related Substances/Degradants, FTIR for ID, Moisture (by Karl Fisher or Loss on Drying), Residual solvents, pH, Weight Variation, Content Uniformity, Friability, Hardness, Disintegration, UV/Visible Spectroscopy, Density or Specific Gravity, Melting Point, Dissolution or Drug Release, Hardness and Particulates.

	Section 2.3.2	Creation, Use, Control, and Maintenance of Equipment Specific Procedures (per instrument, as needed)	The list of equipment procedures can be substantial in that each piece of equipment must have a procedure describing its use.
	Section 2.3.3	Creation, Use, Control, and Maintenance of Product Testing Specifications	This list can be substantial depending upon the number and types of tests required and the number of products being manufactured. Compendial products must meet compendial testing and specification requirements. Some organizations set in-house specifications which may be tighter than compendial or release specifications. These requirements should be addressed in this procedure.
3.0-LE		**SOPs Related to the Laboratory Equipment Sub Element**	This sub element includes all simple and complex instruments and pieces of equipment which are used in or supports testing. See Chapter 4: Components of the Laboratory Equipment Sub Element for additional detail.
SOP-LE-3.1		*Sourcing and Purchasing Laboratory Equipment*	
SOP-LE-3.2		*Guidelines for Equipment Installation, Operation and Performance Qualification (IQ/OQ/PQ)*	IQ/OQ/PQ should follow the guidance given in the current USP <1058> “ANALYTICAL INSTRUMENT QUALIFICATION”.
SOP-LE-3.3		*Instrument Lockout/Tagout Procedure*	
SOP-LE-3.4		*Instrument Calibration and Preventative Maintenance Program*	This should be a description of the calibration and maintenance program.
SOP-LE-3.5		*Instrument Calibration Procedures (per instrument, as needed)*	This is a series of procedures which are written specifically for each piece of equipment or instrument which requires calibration.
SOP-LE-3.6		*Instrument Repair and Preventive Maintenance Procedures (per instrument, as needed)*	This is a series of procedures which are written specifically for each piece of equipment or instrument which requires routine and preventive maintenance.
SOP-LE-3.7		*Calibration Weight Sets*	
SOP-LE-3.8		*UV/Vis Calibration Standards*	
SOP-LE-3.9		*FTIR Calibration Standards*	
SOP-LE-3.10		*Daily Balance Checks*	

(Continued)

Table 3.2 (Continued)

SOP #	Section #	Title or header	Notes and clarifications
4.0-LF		**SOPs Related to the Laboratory Facilities Sub Element**	See Chapter 5: Components of the Laboratory Facilities Sub Element for additional detail.
SOP-LF-4.1		*Laboratory Building and Facility Maintenance*	
SOP-LF-4.2		*Laboratory Water Systems*	The laboratory water system is a critical component to the operation of the QC laboratory. The laboratory water system must be qualified, but not to the extent that the manufacturing water system must be validated. Some laboratories do not have water systems but purchase laboratory water as a reagent.
SOP-LF-4.3		*Laboratory Building and Facilities Security and Access Control*	Laboratory spaces, including workspaces, sample storage areas, office areas, document control rooms, etc. must have limited access. Persons not working with or associated with the laboratory must be escorted.
SOP-LF-4.4		*Temperature and Humidity Monitoring Systems and Chart Recorders*	
SOP-LF-4.5		*Fume Hood Use, Cleaning, and Maintenance*	
SOP-LF-4.6		*Disaster Recovery Plan*	Disaster recovery involves a set of tools and procedures to enable the recovery or continuation of operation of the laboratory following a natural or human-induced disaster. Disaster recovery plans should include periodic transfer and storage of data and records to an offsite location which is not subject to the same potential hazards as the primary data and records storage area. Disaster recovery plans should be rehearsed to ensure they can be successfully implemented in the time of real need.

5.0-MV	**SOPs Related to the Methods Validation and Method Transfer Sub Element**	See Chapter 6: Components of the Method Validation and Method Transfer Sub Element for additional detail.
SOP-MV-5.1	*Analytical Test Method Development*	
SOP-MV-5.2	*Analytical Test Method Validation*	
SOP-MV-5.3	*Analytical Test Method Transfer*	See USP <1224> TRANSFER OF ANALYTICAL PROCEDURES
SOP-MV-5.4	*Qualification of Compendial Test Methods*	See USP <1225> VALIDATION OF COMPENDIAL PROCEDURES
6.0-LC	**SOPs Related to the Laboratory Computers Sub Element**	See Chapter 7: Components of the Laboratory Computer Systems Sub Element for additional detail.
SOP-LC-6.1	*Electronic Records and Signatures*	
SOP-LC-6.2	*Electronic Records Storage, Backup, Archival, and Restoration*	
SOP-LC-6.3	*General Procedures for Computer System Validation*	
SOP-LC-6.4	*Chromatographic Data Acquisition Software*	
SOP-LC-6.5	*Electronic Laboratory Notebooks (ELN)*	
SOP-LC-6.6	*Laboratory Information Management System (LIMS)*	
SOP-LC-6.7	*Computer System Change Control Procedures*	
SOP-LC-6.8	*Validation of Spreadsheets*	
SOP-LC-6.9	*Validation of Databases*	
SOP-LC-6.10	*Power Failure Recovery Procedures for Computers*	
SOP-LC-6.11	*Disaster Recovery of Electronic Data and Computer Equipment*	
SOP-LC-6.12	*Computer System Security*	
SOP-LC-6.13	*Operational Maintenance of Computer Systems and Software*	

(Continued)

Table 3.2 (Continued)

SOP #	Section #	Title or header	Notes and clarifications
7.0-LI		**SOPs Related to the Laboratory Investigations Sub Element**	See Chapter 8: Components of the Laboratory Investigations Sub Element for additional detail.
SOP-LI-7.1		*Conducting, Documenting, and Reporting Laboratory Investigations (LIRs), Out-of-Specification (OOS) and Out-of-Trend Investigations (OOT)*	This procedure can also address deviations (e.g. deviating in an unplanned or planned manner from existing procedures) or refer to the Site procedure regarding deviation investigations.
SOP-LI-7.2		*Procedures for Addressing Unknown or Extraneous Chromatographic Peaks*	
SOP-LI-7.3		*System Suitability Failures: Capture, Tracking and Patterning*	Establishment of system suitability is required for HPLC, GC, and other analyses. System suitability parameters are determined from method validation data. System suitability parameters must be calculated and meet predetermined values prior to actual sample analysis. Failures or repeated failures may indicate method, equipment or personnel problems. Therefore, capture, tracking and patterning of the data can be very useful in improving the efficiency and ruggedness of testing.
8.0 DI		**SOPs Related to Data Governance and Data Integrity Sub Element**	See Chapter 9: Components of the Data Governance and Data Integrity Sub Element for additional detail.
SOP-DI-8.1		*Overview of Data Governance and Data Integrity*	
SOP-DI-8.2		*Data Sources and Data Mapping*	
SOP-DI-8.3		*Application of Hazard Analysis Critical Control Points (HACCP) to Laboratory Data Integrity*	
SOP-DI-8.4		*Personnel Compliance Program for Insuring Laboratory Data Integrity*	

9.0-SB	**SOPs Related to the Stability Program Sub Element**	See Chapter 10: Components of the Stability Program Sub Element for additional detail.
SOP-SB-9.1	*An Overview of the Stability Program*	
SOP-SB-9.2	*Initiating Stability Studies (Study Requests and Protocol Generation)*	
SOP-SB-9.3	*Activating Stability Studies (Sample Set-Down)*	
SOP-SB-9.4	*Time Zero Testing, Review, and Reporting*	
SOP-SB-9.5	*Withdrawal of Drug Product from Stability Chambers (Sample Pull)*	
SOP-SB-9.6	*Stability Sample Testing and Results Reporting*	
SOP-SB-9.7	*Stability Data Analysis, Reporting, and Shelf-Life Assignment*	See Chapter 10 for additional detail regarding expiration dating, initial shelf-life assignment and extension shelf-life and/or retest periods.
SOP-SB-9.8	*Administration of the Stability Program and Stability Studies*	
SOP-SB-9.9	*Additional Types of Stability Studies: Shipping Studies, Thermal Excursion Studies, Bulk and Blend Hold Time Studies, Post Approval Change Studies, and Reserve Samples*	
SOP-SB-9.10	*Selection, Qualification, Calibration, and Maintenance of Stability Chambers*	
10.0-CP	**SOPs Related to the General Laboratory Compliance Practices Sub Element**	See Chapter 11: Components of the General Laboratory Compliance Practices Sub Element for additional detail.
SOP-CP-10.1	*Laboratory Housekeeping and General Appearance*	
SOP-CP-10.2	*Periodic Laboratory-Led CGMP Compliance Inspections*	

Tools and Templates

The following are provided in electronic format as Example Tools and Templates, which can be found in the Chapter 3 Appendix (www.wiley.com/go/Bliesner/LabControl_GMPEnvironment):

- Table 3.1, Model Master Laboratory Control System SOP Index in MS Word® format
- Table 3.1, Model Master Laboratory Control System SOP Index in and Excel® format with additional detail
- An Example-Template for a Standard Operating Procedure
- An Example-Template for a Standard Test Method

4

Components of the Laboratory Equipment Sub Element (LE)

Description of the Laboratory Equipment Sub Element

Since the vast majority of data in a modern analytical laboratory are acquired via instrumentation, it is essential that these instruments have been properly sourced, installed, qualified, and calibrated (were appropriate) and are kept in proper working order. The Laboratory Equipment sub element addresses these issues. It should be noted, that for the purposes of this chapter, the term “equipment” is used interchangeable with the word “instrument” or “instrumentation.”

Typically, the Laboratory Equipment sub element includes at least three individual topics. A laboratory, which is in compliance with Current Good Manufacturing Practices (CGMPs), should address each of these topics (as applicable). These topics include (i) Administrative Procedures Related to the Laboratory Equipment Sub Element Program, (ii) Equipment Specific Standard Operating Procedures (e.g. procedures related to qualification, operation, maintenance, repair, and calibration of each individual type of instrument), and (iii) Laboratory Glassware.

Laboratory glassware is listed as stand-alone topic in that the proper selection, cleaning, care, and use of laboratory glassware is critical to obtaining accurate and consistent results in the quality control (QC) laboratory. In addition, experience has shown that a substantial number of laboratory and Out-of-Specification investigations end up being glassware-related.

These three topics are discussed within the remainder of the text of this chapter along with some suggestions on what items should be addressed in each topic. The list of items under each topic is fairly comprehensive but may not be totally inclusive of all those components, which may constitute the sub element.

Laboratory Control System Operations in a GMP Environment, First Edition. David M. Bliesner.

Companion website: www.wiley.com/go/Bliesner/LabControl_GMPEnvironment

Contents of the Sub Element

A list of example Standard Operating Procedures (SOPs), which outlines the laboratory equipment program, is shown in the succeeding text. These SOPs when fully developed and implemented should describe and instruct how to properly source, install, qualify, calibrate, maintain, and repair laboratory equipment.

Model Standard Operating Procedures for Establishing and Maintaining an Effective Laboratory Equipment Program

1. **Procedure #1: Overview of the Laboratory Equipment Program**
   - 1.1 Program Description
   - 1.2 Roles and Responsibilities
   - 1.3 Training and Qualification
   - 1.4 Equipment Vendor Contracts and Relationship Management
2. **Procedure #2: Assessing Laboratory Equipment Needs**
   - 2.1 Description of samples currently tested in the laboratory
   - 2.2 Description tests and associated equipment for each test
   - 2.3 Appropriateness of equipment for current testing needs
   - 2.4 Assessment of future types of samples to be tested in the laboratory
   - 2.5 Assessment of workload projections, future testing requirements and associated equipment needs
3. **Procedure #3: Sourcing and Purchasing Laboratory Equipment**
   - 3.1 Executing a needs assessment
   - 3.2 List of approved vendors
   - 3.3 Capital equipment budget
   - 3.4 Planned and scheduled purchases
   - 3.5 Planned and scheduled replacements
   - 3.6 Unscheduled replacements
4. **Procedure #4: Laboratory Equipment Inventory, Inventory Control and Usage Tracking**
   - 4.1 Controlled-access master equipment lists
   - 4.2 Master equipment maps
   - 4.3 Controlled access master equipment records
   - 4.4 Equipment and component labeling
   - 4.5 Equipment usage logs
   - 4.6 Instrument lockout/tagout procedures
5. **Procedure #5: Installation, Operational, and Performance Qualification (IQ/OQ/PQ) of Laboratory Equipment**
   - 5.1 Alignment with USP <1058> ANALYTICAL INSTRUMENT QUALIFICATION
   - 5.2 New equipment qualifications

   5.3 Re-Qualifications
   5.4 Documentation
   5.5 Design Qualification
6. **Procedure #6: Equipment Maintenance and Calibration**
   6.1 Alignment with equipment manufacturers
   6.2 Maintenance schedules
   6.3 Maintenance records
   6.4 Calibration requirements
   6.5 Calibration records
   6.6 Failed calibrations and investigations
7. **Procedure #7: Repairing Laboratory Equipment**
   7.1 Repair requests
   7.2 Performance of repairs
   7.3 Repair records
   7.4 Parts lockers
8. **Procedure #8: Laboratory Equipment Lifecycle Management**
   8.1 Definition of equipment lifecycle management
   8.2 Capturing and tracking equipment related metrics
   8.3 Analyzing equipment related metrics
   8.4 Reporting equipment related metrics
   8.5 Decommissioning and replacement of laboratory equipment
   8.6 Retention of critical components
   8.7 Like-for-like exchanges
   8.8 Non-like purchases
9. **Procedure #9: United States Pharmacopeia (USP) Instrument Groupings and Equipment Related Guidance Chapters**
   9.1 Description of USP <1058> ANALYTICAL INSTRUMENT QUALIFICATION instrument groupings
   9.2 USP <1058> Group A
   9.3 USP <1058> Group B
   9.4 USP <1058> Group C
   9.5 Equipment related USP guidance chapters
10. **Procedure #10: Laboratory Glassware**
   10.1 Importance of laboratory glassware
   10.2 Types and uses of laboratory glassware
   10.3 Sourcing, purchasing and receipt of laboratory glassware
   10.4 Cleaning of laboratory glassware
   10.5 Storage and maintenance of laboratory glassware
   10.6 Disposal of laboratory glassware

The model SOP list earlier is described in additional detail, along with some notes and clarifications in Table 4.1 “Components of the laboratory equipment program." The list of items under each topic is fairly comprehensive but may not be totally inclusive of all those components, which may constitute the sub element. Individual organizational requirements will vary.

Table 4.1 Components of the laboratory equipment program.

#	Laboratory equipment sub element topic	Notes and clarifications
4.1	***Administrative and Operational Procedures Related to the Laboratory Equipment Sub Element***	
4.1.1	**Procedure #1**: Overview of the Laboratory Equipment Program	This procedure describes the major operational components of the Laboratory Equipment program (e.g. sub element).
4.1.1.1	Program Description	This section should describe the purpose of the laboratory equipment program, define some key terms, and discuss linkage to the Site Master Validation Plan.
4.1.1.2	Roles and Responsibilities	This section should list positions within the organizational structure in the lab who are part of the laboratory equipment program including: • Supervisors/Managers • Equipment Support Technicians/Metrologists • Analysts • Vendors This section should also list and discuss who is responsible for specific duties related to the laboratory equipment program such as: • Compiling and maintaining master equipment lists • Establishing and maintain parts lockers • Creating maintenance schedules • Creating calibration schedules • Re-qualification schedules • Requisition of new equipment • Etc.

(Continued)

Table 4.1 (Continued)

#	Laboratory equipment sub element topic	Notes and clarifications
4.1.1.3	Training and Qualification	Training is defined as an activity designed to provide the skills and/or knowledge that individuals need in order to perform assigned job-related duties and responsibilities. Qualification is defined as the process, including documentation and approval, whereby a person is trained and demonstrates proof-of-proficiency in specified tests, tasks, procedures, or techniques. All personnel associated with the laboratory equipment program must possess the appropriate education, training, and experience in the field of laboratory equipment to be assigned duties within this sub element. This should be documented in their training records.
4.1.1.4	Equipment Vendor Contracts and Relationship Management	Use of vendors to help support the laboratory equipment program is common. It is important to spell out how these vendors are integrated into the laboratory equipment program via implementation of a Quality Agreement. This agreement should address at least the following (as appropriate): • Equipment qualification and re-qualification • Scheduled maintenance • Repairs • Calibration • Maintenance of parts lockers • Provide equipment related training and best practices training to laboratory personnel • Other services as appropriate
4.1.1.5	Specialized Sample Analysis	Small organizations may not have the resources, or sample testing volume to justify purchase, qualification, maintenance, and calibration of certain types of equipment (e.g. NMR, X-Ray, etc.). In these cases, samples may be sent to an outside vendor to perform the testing. These relationships must be carefully managed, and in some cases even more so than your on-site vendors.

(Continued)

Table 4.1 (Continued)

#	Laboratory equipment sub element topic	Notes and clarifications
4.1.2	**Procedure #2**: Assessing Laboratory Equipment Needs	An initial and periodic re-assessment of the laboratory's equipment needs should be performed. This will ensure that there is a proper balance between testing needs and availability of equipment to support the testing.
4.1.2.1	Description of Samples Currently Tested	
4.1.2.2	Description of Tests and Associated Equipment for Each Test Such as: • Description; Visual within Light Box • Assay; HPLC, Titration, GC • Impurities/Related Substances/Degradants; HPLC, GC • Identification – FTIR, ultra-violet (UV), HPLC • Moisture; Karl Fisher Titration, Loss on Drying, Thermal Gravimetric Analysis (TGA) • Residual Solvents; GC Headspace • pH; pH Meter • Weight Variation; Analytical Balance • Content Uniformity; HPLC • Friability; Friabilator • Hardness; Hardness Tester • Disintegration; Disintegration Tester • Density or Specific Gravity; Density Meter, Specific Gravity Meter • Melting Point; Melting Point Apparatus • Dissolution or Drug Release; Dissolution Apparatus • Particulates; Particle Tester • Optical Rotation; Polarimeter • Viscometer (rotational) • Osmometer	Laboratory management needs to be aware of all the types of samples, testing performed on the samples, and the equipment used to perform the testing. Some examples are listed in the succeeding text.

(Continued)

Table 4.1 (Continued)

#	Laboratory equipment sub element topic	Notes and clarifications
4.1.2.3	Appropriateness of Equipment for Current Testing Needs	Better methods, techniques and equipment, and automation may become available over time. This evaluation will insure compliant and efficient laboratory operations. The serviceability (e.g. does it function properly and consistently) of existing equipment should also be made on a periodic basis.
4.1.2.4	Assessment of Future Types of Samples to be Tested in the Laboratory	This allows for planned purchases, retirement, or decommissioning of equipment.
4.1.2.5	Assessment of Workload Projections, Future Testing Requirements and Associated Equipment Needs	This will ensure a balance exists between available personnel, types, and quantities of equipment and sample testing throughput.
4.1.3	**Procedure #3**: Sourcing and Purchasing Laboratory Equipment	
4.1.3.1	Executing a Needs Assessment	As before, this will ensure a balance exists between available personnel, types, and quantities of equipment and sample testing throughput.
4.1.3.2	List of Approved Vendors	The Quality Assurance (QA) group maintains an approved vendor list. If the list does not contain a vendor who supplies the equipment type you are looking for, or you are sourcing secondary vendors, you will need to follow your site procedure for qualifying vendors and executing a quality agreement with the new vendors. See FDA guidance document titled "Contract Manufacturing Arrangements for Drugs: Quality Agreements" November 2016 for additional information.
4.1.3.3	Capital Equipment Budget	The laboratory should have at least three separate budgets. One for operational items and supplies, one for training and qualification of personnel and one for purchasing equipment and instruments (e.g. the capital equipment budget).
4.1.3.4	Planned and Scheduled Purchases	

(Continued)

Table 4.1 (Continued)

#	Laboratory equipment sub element topic	Notes and clarifications
4.1.3.5	Planned and Scheduled Replacements	The laboratory must plan and schedule for replacing equipment. Unfortunately, it is a common practice, not to plan ahead often leaving the laboratory using instruments which are marginally serviceable.
4.1.3.6	Unplanned Replacements	Unplanned replacements can be the result of unrepairable equipment failures or the requirement for new types of testing.
4.1.4	**Procedure #4**: Laboratory Equipment Inventory, Inventory Control and Usage Tracking	
4.1.4.1	Controlled Access Master Equipment Lists	This can take the form of proprietary software (e.g. database), controlled spreadsheets, paper records or hybrid paper, and electronic records. Access to this list should be restricted to ensure its accuracy.
4.1.4.2	Master Equipment Maps	Master equipment maps show the current location of all equipment in all laboratories. This can take the form of proprietary software (e.g. database), controlled spreadsheets, paper records or hybrid paper, and electronic records. For very large laboratories with numerous equipment assets, radio-frequency identification (RFID), or "smart label" tracking may be employed.
4.1.4.3	Controlled Access Master Equipment Records	These are the living history for each piece of equipment and should include records related to purchase, qualification, maintenance, calibration, repairs, re-qualification, and decommissioning. They can take the form of proprietary software (e.g. database), controlled spreadsheets, paper records or hybrid paper, and electronic records. Access to these records list should be restricted to ensure the integrity of the equipment records.

(Continued)

Table 4.1 (Continued)

#	Laboratory equipment sub element topic	Notes and clarifications
4.1.4.4	Equipment and Component Labeling	Each piece of equipment or component (as in the case of HPLCs, which consist of multiple modules working together) need to be labeled. The label may include asset numbers, asset names, qualification dates, calibration dates, recalibration/ requalification dates, etc.
4.1.4.5	Equipment Usage Logs	The good manufacturing practices (GMPs) in Section 211.182 "Equipment cleaning and use log, requires that each piece of equipment have an associated usage log." This log is a record of the analyses performed, some explanation of the types of testing performed and documentation of any instrument issues. Although not specifically required for laboratory equipment, evidence of cleaning may be documented in this log as well.
4.1.4.6	Instrument Lockout/Tagout Procedures	These procedures prevent unauthorized use of equipment in the event it needs to be removed from service for any reason.
4.1.5	**Procedure #5**: Installation, Operational and Performance Qualification (IQ/OQ/PQ) of Laboratory Equipment	
4.1.5.1	Alignment with USP <1058> ANALYTICAL INSTRUMENT QUALIFICATION	USP <1058> is a clear, concise, and practical guide to performing equipment qualification. This should lay the basis for any instrument qualification SOPs.
4.1.5.2	New Equipment Qualifications	IQ/OQ/PQ exercises are protocol driven with predetermined acceptance criteria. Criteria should be based on vendor recommendations and scientific principles.
4.1.5.3	Re-Qualifications	Re-qualifications are triggered for two major reasons including (i) time-based (e.g. every so many months, once a year, etc.) and (ii) event-based (e.g. following equipment relocation, preventive maintenance, major repairs, etc.).

(Continued)

Table 4.1 (Continued)

#	Laboratory equipment sub element topic	Notes and clarifications
4.1.5.4	Documentation	The complete IQ/QQ/PQ documentation package is retained in the permanent equipment record file. The package should include executed protocol, which has been reviewed and approved by QA, and may include any supporting instrument outputs, copies of logbook pages, etc. It is perfectly acceptable to use pre-existing Vendor templates as long as they have been reviewed for accuracy and completeness and have site QA review and approval.
4.1.5.5	Design Qualification (DQ)	According to USP <1058> Design Qualification (DQ) is the documented collection of activities that define the functional and operational specifications and intended purpose of the instrument. DQ may (and usually is) performed by the instrument manufacturer or the user. It is expected that DQ requirements will be minimal for commercial, off-the-shelf instruments. Verification that the instrument specifications meet the desired functional requirements may suffice.
4.1.6	**Procedure #6:** Equipment Maintenance and Calibration	Equipment maintenance and calibration requirements are instrument specific.
4.1.6.1	Alignment with Equipment Manufacturers	Any maintenance or calibration procedures should be generated by first reviewing the manufacturer's literature and equipment manuals. In many cases there will be minimal need to create these procedures "from scratch."
4.1.6.2	Maintenance Schedules	Maintenance schedules and their corresponding actions are created using the manufacturer's recommendations as well as the laboratory's own experience with the equipment. For example, HPLC injector rotor seals may need more frequent changing if the lab analyzes samples which have a tendency to contaminate and/or wear the seals more than usual.
4.1.6.3	Maintenance Records	Maintenance records are stored as part of the master equipment record.

(Continued)

Table 4.1 (Continued)

#	Laboratory equipment sub element topic	Notes and clarifications
4.1.6.4	Calibration Requirements	Same as for maintenance. Follow manufacturer's recommendations and augment as needed.
4.1.6.5	Calibration Records	Same as for maintenance.
4.1.6.6	Failed Calibration and Investigations	Calibration exercises usually require documentation of "as found" and "as left." So that any trends in instrument performance are captured and tracked. In the event the instrument is found out of calibration at the beginning of the exercise, a formal laboratory investigation needs to be performed to assess the impact of the failure on any testing results since the last calibration exercise.
4.1.7	**Procedure #7**: Repairing Laboratory Equipment	The process of repairing laboratory equipment should be captured in a work flow diagram within this procedure.
4.1.7.1	Repair Requests	Repair requests are formal requests, which may be paper, electronic, or combination of both (e.g. a hybrid system). Repair requests should be tracked from opening to closure and become part of the instruments master equipment record, or the record should have linkage to the completed request (e.g. if electronic).
4.1.7.2	Performance of Repairs	If laboratory personnel are trained and qualified to perform equipment repairs this should be highlighted in the document. Also, the steps needed for the equipment vendors to be tasked to perform the repairs needs to be described in this procedure.
4.1.7.3	Repair Records	As mentioned previously, repair records become part, or at least a linkage to, the master equipment record.
4.1.7.4	Parts Lockers	In order to speed up the process of equipment repairs, it is sometimes useful to maintain an equipment parts locker on site. The parts locker contains the most common replacement parts, and may also include parts, which may take a long time to receive if ordered at the time of equipment failure. Parts lockers can be maintained by laboratory personnel or by the vendor who performs repairs.

(Continued)

Table 4.1 (Continued)

#	Laboratory equipment sub element topic	Notes and clarifications
4.1.8	**Procedure #8**: Laboratory Equipment Lifecycle Management	
4.1.8.1	Definition of equipment lifecycle management	Equipment lifecycle is defined as all the phases in the life of the equipment including, but not limited to sourcing, purchase, qualification, maintenance, calibration, repairs, re-qualification, and decommissioning.
4.1.8.2	Capturing and Tracking Equipment Related Metrics	Metrics related to testing failures (with linkage to instrumentation used during the testing) and equipment related failures (instrument issues which resulted in instrument faults, failure of system suitability, testing failures, etc.) should be captured and tracked for periodic for analysis.
4.1.8.3	Analyzing Equipment Related Metrics	Equipment related metrics can be patterned using tools such as Pareto charts, clustered bar charts, pie charts, or other techniques, which allow visualization of the data. By linking the observations to the type of instrument (e.g. HPLC, UPLC, GC) and specific instrument itself (e.g. instrument identified by equipment number), patterns may emerge from the data. For example, by patterning failures by method, by instrument, by type of failure, etc. it may be possible to show that certain pieces of equipment have reached the end of their serviceable life.
4.1.8.4	Reporting Equipment Related Metrics	This is most commonly performed at laboratory quality review meetings and/or QA quality review meetings.
4.1.8.5	Decommissioning and Replacement of Laboratory Equipment	Decommissioning and replacing equipment should be done in a systematic and fact-based manner, using the metrics previously mentioned.

(Continued)

Table 4.1 (Continued)

#	Laboratory equipment sub element topic	Notes and clarifications
4.1.8.6	Retention of Critical Components	A risk assessment needs to be performed to ensure that data collected and reported in the past on decommissioned systems can be accessed and processed, and that the results and reports can be confirmed or recreated if necessary. This may require retention of the hardware, computer systems, and software.
4.1.8.7	Like-for-Like Exchanges	A risk assessment which includes subject matter experts and QA personnel needs to be performed to insure that "like-for-like" equipment exchanges are in fact valid.
4.1.8.8	Non-Like Exchanges and Purchases	Cross-over studies, requalification, or even method revalidations maybe be required to implement non-like equipment exchanges.
4.1.9	**Procedure #9**: United States Pharmacopeia (USP) Instrument Groupings and Equipment Related Guidance Chapters	
4.1.9.1	Description of USP <1058> ANALYTICAL INSTRUMENT QUALIFICATION instrument groupings	The USP groups instruments (equipment) into three separate categories (A, B, and C) based on complexity.
4.1.9.1.1	USP <1058> Group A	• Least complex instruments • No measurement capability • Qualification = observation of proper function • Examples: Magnetic stirrer, vortex mixer
4.1.9.1.2	USP <1058> Group B	• May provide a measurement or an experimental condition that can affect a measurement • Have measurement values requiring calibration • Qualification = observation of proper function of instruments and may require only routine calibration, maintenance, or performance checks • Examples: pH meter, oven

(Continued)

Table 4.1 (Continued)

#	Laboratory equipment sub element topic	Notes and clarifications
4.1.9.2	USP <1058> Group C	• Analytical instruments with a significant degree of computerization and complexity. • Qualification = All elements of qualification including software validation must be considered to ensure proper functioning of instruments in this group. • Examples: HPLC, GC, mass spectrometers.
4.1.9.3	Equipment Related USP Guidances	The USP provides numerous equipment related guidances, which address theory and practice related to the equipment. Examples include USP <429> LIGHT DIFFRACTION MEASUREMENT OF PARTICLE SIZE, and USP <854> MID-INFRARED SPECTROSCOPY, which both discuss instrument qualification. Table 4.2 shows additional examples.
4.2	***Equipment Specific Standard Operating Procedures***	
	See Table 4.2 "Suggestions for Laboratory Equipment-Related Standard Operating Procedures," which follows Section 4.3 *Laboratory Glassware*, for examples of the content for equipment specific procedures.	Table 4.2 below titled "Suggestions for Laboratory Equipment-Related Standard Operating Procedures" is designed to provide a starting point for creating site-specific, instrument-specific procedures for (i) qualifying equipment, (ii) operating equipment, (iii) maintaining equipment, and (iv) repairing equipment (as necessary) and calibrating equipment (as appropriate). The numbers and types of procedures will depend on the types of testing and types of equipment that exist in a particular laboratory.

(Continued)

Table 4.1 (Continued)

#	Laboratory equipment sub element topic	Notes and clarifications
4.3	***Laboratory Glassware***	
4.3.1	**Procedure #10**: Laboratory Glassware	
4.3.1.1	Importance of Glassware	Glassware is a ubiquitous component of all QC and R&D laboratories. It is used to prepare sample solutions, standard solutions, mobile phases, volumetric solutions, and has additional uses. Because of its ubiquitous nature, it is important to understand how to properly choose, use, and care for laboratory glassware.
4.3.1.2	Types and Uses of Laboratory Glassware	Types of glassware used in the laboratory include beakers, flasks, graduated cylinders, pipets, Erlenmeyer flasks, and burets. As stated above, their uses include preparation of sample solutions, standard solutions, mobile phases, volumetric solutions, as weighing vessels, in titrations, etc. Laboratory personnel should be trained on the proper use of all types of laboratory glassware.
4.3.1.3	Sourcing, Purchasing and Receipt of Laboratory Glassware	Laboratory glassware can be purchased from any reputable laboratory supply house. When purchasing glassware, it is mandatory that the purchased pieces are correct glassware for their intended use including the Class (A or B), and function (to contain [TC] or to deliver [TD]) for pipets.

(Continued)

Table 4.1 (Continued)

#	Laboratory equipment sub element topic	Notes and clarifications
4.3.1.4	Cleaning of Laboratory Glassware	Glassware can be cleaned by using automated glassware washers or by manual means. Regardless of technique, the glassware washing must be validated to ensure the removal of chemical residue. Validation may take the form of "dirtying" the glassware with a worse-case material, cleaning it, and checking for residue. Laboratory personnel should be adept at identifying unclean glassware in that dirty glassware can lead to contaminated standards and samples, which may result in unknown peaks during chromatographic analyses.
4.3.1.5	Storage and Maintenance of Laboratory Glassware	Glassware should be stored as recommended by the manufacturer. This may include placing it in sliding drawers in holding trays designed specifically for each type of glassware. Glassware used to prepare basic solutions and sample should be scrutinized more carefully for wear and serviceability in that bases degrade glassware over time. Routine, preventive maintenance of the automated glassware washer should be delineated in a written procedure. Consider using a Polariscope for periodic inspection of glassware to assess its serviceability (e.g. identifying defective and worn glassware) by observing stress lines or fractures in the glass which appear over time.
4.3.1.6	Disposal of Laboratory Glassware	Glassware should be checked before use and periodically inventoried and inspected for serviceability. If glassware serviceability is suspect-dispose of it in a broken glassware receptacle.

Table 4.2 Suggestions for laboratory equipment-specific Standard Operating Procedures.

Equipment or instrument	USP <1058> category	Typical testing applications	Suggestions for laboratory equipment-related SOPs
HPLC or UPLC	Group C: Complex Instruments. Requires full qualification and specific function and performance tests.	Assay, impurities, content uniformity, identification (retention time match)	It is difficult to create operational procedures for complex instruments like HPLCs and UPLCs (ultra-performance chromatography). The best approach is to train and qualify (e.g. demonstrated proof-of-proficiency) analysts and have them follow specific test methods after the training and qualification. In general, the chromatography skills required to perform UPLC analyses, is greater than HPLC analyses. In particular, extra care is needed to keep mobile phases, reference standard solutions, and samples free of particulate matter. Also, equipment needs to be scrupulously maintained. This includes paying particular attention to dirty solvent filters, unclean check valves, improperly cut tubing, unclean or worn piston seals, and plungers. HPLC and UPLCs methods may not be interchangeable due to the higher pressures and smaller volume plumbing associated with UPLC. If the method was not validated for both HPLC and UPLC, some assessment needs to be made. The following equipment related procedures should be considered for liquid chromatographs: • SOP addressing IQ/OQ/PQ for liquid chromatographic systems • System repair and maintenance: Pre-run and post run checks, preventative maintenance, and repairs for cause • Periodic, scheduled re-qualification, or re-qualification for cause (e.g. relocation of systems, major repairs, module swap-out, etc.) Note that USP <621> CHROMATOGRAPHY, provides additional information and guidance with respect to procedures for HPLC, UPLC, and GC.

(Continued)

Table 4.2 (Continued)

Equipment or instrument	USP <1058> category	Typical testing applications	Suggestions for laboratory equipment-related SOPs
GC	Group C: Complex Instruments. Requires full qualification and specific function and performance tests.	Assay, impurities of volatile materials	In general, the chromatography skills required for GC are different than HPLC or UPLC. Personnel performing GC analyses should not only be trained to perform GC analyses but they should also be qualified to include proof-of-proficiency similar to liquid chromatography. The following equipment related procedures should be considered for gas versus liquid chromatographs: • SOP addressing IQ/OQ/PQ for gas chromatographic systems • System maintenance: pre-run and post run checks, preventative maintenance, and repairs for cause • Periodic, scheduled re-qualification or re-qualification for cause (e.g. relocation of systems, major repairs, etc.) Note that USP <621> CHROMATOGRAPHY, provides additional information and guidance with respect to procedures GC.
GC Headspace	Group C: Complex Instruments. Requires full qualification and specific function and performance tests.	Residual solvents of active pharmaceutical ingredients (APIs) and finished products (FPs) (where appropriate)	In addition to the points highlighted above for GC, the following equipment related procedures should be considered for gas chromatographs with headspace autosamplers (analyzers): • SOP addressing IQ/OQ/PQ for the headspace autosamplers attached to the gas chromatographic systems • System maintenance: pre-run and post run checks, preventative maintenance, and repairs for cause • Periodic, scheduled re-qualification or re-qualification for cause (e.g. relocation of systems, major repairs, etc.) Note USP <621> CHROMATOGRAPHY, provides additional information and guidance with respect to procedures for HPLC, UPLC, and GC. In addition, USP <467> RESIDUAL SOLVENTS and <1467> RESIDUAL SOLVENTS VERIFICATION OF COMPENDIAL PROCEDURES AND VALIDATION OF ALTERNATIVE PROCEDURES should be referenced when creating site-specific SOPs related to headspace analyses.

UV/Visible Spectroscopy	Group C: Complex Instruments. Requires full qualification and specific function and performance tests.	Identification testing, dissolution testing	The following equipment related references, along with the instrument manufacturer's operating manuals and supporting literature, should be used as a basis for creating your site-specific procedures for qualifying, operating, maintaining, repairing, and calibrating ultra-violet/visible (UV/Vis) spectrometers: • USP <857> ULTRAVIOLET-VISIBLE SPECTROSCOPY • USP <1857> ULTRAVIOLET-VISIBLE SPECTROSCOPY-THEORY AND PRACTICE
FTIR Spectroscopy	Group C: Complex Instruments. Requires full qualification and specific function and performance tests.	Identification testing	The following equipment related references, along with the instrument manufacturer's operating manuals and supporting literature, should be used as a basis for creating your site-specific procedures for qualifying, operating, maintaining, repairing, and calibrating FTIR spectrometers: • USP <854> MID-INFRARED SPECTROSCOPY • USP <1854> MID-INFRARED SPECTROSCOPY-THEORY AND PRACTICE
Karl Fischer Titrator	Group B: Standard instruments that have measurement values requiring calibration.	Moisture analysis	The following equipment related references, along with the instrument manufacturer's operating manuals and supporting literature, should be used as a basis for creating your site-specific procedures for qualifying, operating, maintaining, repairing, and calibrating Karl Fischer titrators: • USP <921>WATER DETERMINATION
Thermal Gravimetric Analysis (TGA)	Group C: Complex Instruments. Requires full qualification and specific function and performance tests.	TGA is used to investigate dehydration/desolvation processes and compound decomposition.	The following equipment related references, along with the instrument manufacturer's operating manuals and supporting literature, should be used as a basis for creating your site-specific procedures for qualifying, operating, maintaining, repairing, and calibrating thermal gravimetric analysis analyzers: • USP <891> THERMAL ANALYSIS

(Continued)

Table 4.2 (Continued)

Equipment or instrument	USP <1058> category	Typical testing applications	Suggestions for laboratory equipment-related SOPs
Differential Scanning Calorimeter (DSC)	Group C: Complex Instruments. Requires full qualification and specific function and performance tests.	Identity testing, purity measurements.	The following equipment related references, along with the instrument manufacturer's operating manuals and supporting literature, should be used as a basis for creating your site-specific procedures for qualifying, operating, maintaining, repairing, and calibrating DSCs: • USP <891> THERMAL ANALYSIS
Analytical Balance	Group B: Standard instruments that have measurement values requiring calibration.	Sample preparation, standard preparation, loss on drying, weight loss analysis, etc.	Refer to USP <41> BALANCES which discusses: • Use for substances which must be "accurately weighed" • Repeatability assessment • Assessment of balance accuracy Refer USP <1251> WEIGHING ON AN ANALYTICAL BALANCE which discusses: • Installation (e.g. support surface, and location) • Operational qualification • Performance qualification • Performance checks and acceptance criteria • Balance checks (e.g. daily or before use) • Receivers (e.g. weighing boats, bottles, paper, etc.) • Types of weighing (e.g. addition weighing, weighing by difference) • Problem samples (e.g. samples which become charged with static electricity, warm samples, etc.)

Top Loading Balance	Group B: Standard instruments that have measurement values requiring calibration.	Mobile phase preparation, reagent preparation, etc.	Top loading balances should be qualified and used in accordance with USP <41> BALANCES and USP <1251> WEIGHING ON AN ANALYTICAL BALANCE.
Friabilator	Group A: Least complex where qualification = observation of proper function	Tablet friability testing, which supplements other physical strength measurements, such as tablet breaking force.	Follow USP <1216> TABLET FRIABILITY along with the instrument manufacturer's operating manuals to create your site-specific procedures for operating and maintaining the friabilator.
Tablet Hardness Tester	Group A: Least complex where qualification = observation of proper function.	Tablet hardness testing (tablet breaking force)	Follow USP <1217> TABLET BREAKING FORCE along with the instrument manufacturer's operating manuals to create your site-specific procedures for operating and maintaining the tablet hardness tester.
Disintegration Tester	Group A: Least complex where qualification = observation of proper function.	To determine whether tablets or capsules disintegrate within the prescribed time.	Follow USP <701> DISINTEGRATION along with the instrument manufacturer's operating manuals to create your site-specific procedures for operating and maintaining the disintegration tester.
Density (Specific Gravity) Meter	Group B: Standard instruments that have measurement values requiring calibration.	Density of solutions, specific gravity of solutions.	Follow USP <841> SPECIFIC GRAVITY in the section related to oscillating transducer density meter, along with the instrument manufacturer's operating manual to create your site-specific procedures for operating, calibrating, and maintaining the density meter.

(Continued)

Table 4.2 (Continued)

Equipment or instrument	USP <1058> category	Typical testing applications	Suggestions for laboratory equipment-related SOPs
Melting Point Apparatus	Group B: Standard instruments that have measurement values requiring calibration.	Melting point measurement of APIs and other pure substances.	USP <741> MELTING RANGE OR TEMPERATURE allows apparatuses with cameras or other computerized equipment to improve accuracy, sensitivity, or precision of the measurement of melting points. Use the instrument manufacturer's operating manual to create your site-specific procedures for operating, calibrating and maintaining the device.
Dissolution Apparatus	Group C: Complex Instruments. Requires full qualification and specific function and performance tests.	Dissolution of tablets, capsules, transdermal patches, etc.	The following equipment related references, along with the instrument manufacturer's operating manuals and supporting literature, should be used as a basis for creating your Site-specific procedures for qualifying and operating dissolution apparatuses: • USP <711> DISSOLUTION • USP "Dissolution Toolkit Procedures for Mechanical Calibration and Performance Verification Test Apparatus 1 and Apparatus 2" Version 2.0, 22 March 2010, for IQ/OQ/PQ.
Particle Size Analyzer	Group C: Complex Instruments. Requires full qualification and specific function and performance tests.	Particle size analyses of API and suspensions.	Follow USP <429> LIGHT DIFFRACTION MEASUREMENT OF PARTICLE SIZE, section titled CONTROL OF THE INSTRUMENT PERFORMANCE along with the instrument manufacturer's operating manuals and supporting literature, as a basis for creating your site-specific procedures for qualifying, operating, maintaining, and calibrating the particle size analyzer.
Polarimeter	Group B: Standard instruments that have measurement values requiring calibration.	Confirmation of specific optical rotation for chiral APIs.	Follow USP <781> OPTICAL ROTATION along with the instrument manufacturer's operating manuals and supporting literature, as a basis for creating your site-specific procedures for qualifying, operating, maintaining, and calibrating the polarimeter.

pH Meter	Group B: Standard instruments that have measurement values requiring calibration.	Mobile phase preparation, sample analysis.	Follow USP <791> pH along with the instrument manufacturer's operating manuals and supporting literature, as a basis for creating your site-specific procedures for qualifying, operating, maintaining, and calibrating the pH meter.
Viscometer (Rotational)	Group B: Standard instruments that have measurement values requiring calibration.	Measuring the viscosity of fluids.	Follow USP <912> VISCOSITY-ROTATIONAL METHODS along with the instrument manufacturer's operating manuals and supporting literature, as a basis for creating your site-specific procedures for qualifying, operating, maintaining, and calibrating the viscometer.
Osmometer	Group B: Standard instruments that have measurement values requiring calibration.	For determining whether a parenteral solution is hypo-osmotic, iso-osmotic, or hyperosmotic.	Follow USP <785> OSMOLALITY AND OSMOLARITY along with the instrument manufacturer's operating manuals and supporting literature, as a basis for creating your site-specific procedures for qualifying, operating, maintaining, and calibrating the osmometer.
Particle Counter	Group B: Standard instruments that have measurement values requiring calibration.	To count extraneous undissolved particles, other than gas bubbles, unintentionally present in the solutions.	Follow USP <788> PARTICULATE MATTER IN INJECTIONS along with the instrument manufacturer's operating manuals and supporting literature, as a basis for creating your site-specific procedures for qualifying, operating, maintaining, and calibrating the instruments used to determine particle matter in samples.

Tools and Templates

The following is provided in electronic format as an Example Tool and Template, which can be found in the Chapter 4 Appendix (www.wiley.com/go/Bliesner/LabControl_GMPEnvironment):

- Model list of Standard Operating Procedures for Establishing and Maintaining an Effective Laboratory Equipment Program in MS Word® format.

References

1 Current USP <1058> ANALYTICAL INSTRUMENT QUALIFICATION.
2 Valigra, L. (2010). Qualifying analytical instruments: general chapter <1058> clarifies terminology, classifies instruments. *Qual. Assur. J.* 13: 67–71. https://doi.org/10.1002/qaj.475.
3 USP <41> BALANCES.
4 USP <1251> WEIGHING ON AN ANALYTICAL BALANCE.
5 Cerdanyola del Vallès (2014). Waters Corporation Training Slide Deck for UPLC Troubleshooting and Diagnostics Tips and Tricks Waters Corporation presented by Cerdanyola del Vallès.
6 Restek (2000). *A Technical Guide for Static Headspace Analysis Using GC.* Bellefonte, PA: Restek Corporation, 110 Benner Circle.
7 John C. Travis, Melody V. Smith, Stanley D. Rasberry, and Gary W. Kramer. Technical Specifications for Certification of Spectrophotometric NTRMs. National Institute of Standards and Technology Special Publication 260-140, 59 pages (February 2000) CODEN: NSPUE2. Washington, DC: U.S. Government Printing Office, Internet: bookstore.gpo.gov-Phone: (202) 512-1800- Fax: (202) 512-2250, Mail: Stop.
8 USP <429> LIGHT DIFFRACTION MEASUREMENT OF PARTICLE SIZE.
9 USP <467> RESIDUAL SOLVENTS.
10 USP <1467> RESIDUAL SOLVENTS-VERIFICATION OF COMPENDIAL, PROCEDURES AND VALIDATION OF ALTERNATIVE PROCEDURES.
11 USP <621> CHROMATOGRAPHY.
12 USP <701> DISINTEGRATION.
13 USP <711> DISSOLUTION.
14 USP <731> LOSS ON DRYING.
15 USP <741> MELTING RANGE OR TEMPERATURE.
16 USP <781> OPTICAL ROTATION.
17 USP <785> OSMOLALITY AND OSMOLARITY.
18 USP <788> PARTICULATE MATTER IN INJECTIONS.

19 USP <791> pH.

20 USP <841> SPECIFIC GRAVITY.

21 USP <854> MID-INFRARED SPECTROSCOPY.

22 USP <1854> MID-INFRARED SPECTROSCOPY-THEORY AND PRACTICE.

23 USP <857> ULTRAVIOLET-VISIBLE SPECTROSCOPY.

24 USP <1857> ULTRAVIOLET-VISIBLE SPECTROSCOPY-THEORY AND PRACTICE.

25 USP <891> THERMAL ANALYSIS.

26 USP <912> VISCOSITY-ROTATIONAL METHODS.

27 USP <921> WATER DETERMINATION.

28 USP <1094> CAPSULES—DISSOLUTION TESTING AND RELATED QUALITY ATTRIBUTES.

29 USP <1216> TABLET FRIABILITY.

30 USP <1217> TABLET BREAKING FORCE.

31 USP (2010). Dissolution Toolkit Procedures for Mechanical Calibration and Performance Verification Test Apparatus 1 and Apparatus 2, Version 2.0, 22 March 2010.

32 FDA (2016). Guidance Document, Contract Manufacturing Arrangements for Drugs: Quality Agreements. November 2016.

33 ICH Harmonised Tripartite Guideline Quality Risk Management Q9, 09 November 2005.

34 Miller, J.M. and Crowther, J.B. (eds.) (2000). *Analytical Chemistry in a GMP Environment: A Practical Guide*, 110–111. New York: Wiley.

35 Shugar, G.J. and Ballinger, J.T. (1996). *Chemical Technicians' Ready Reference Handbook*, 4e, 339–368. New York: McGraw-Hill, Inc.

5

Components of the Laboratory Facilities Sub Element (LF)

Description of the Laboratory Facilities Sub Element

As with a manufacturing area, it is essential that the laboratory be properly designed and equipped. This ensures that the data generated accurately reflect the true composition of the sample from which they came. In addition, it ensures a safe working environment for laboratory personnel.

Laboratory facilities include four individual topics. A laboratory, which is in compliance with CGMPs, should at least address each of these topics (as applicable). These topics include (i) Physical Construction and Layout of the Laboratory, (ii) Utilities and Services, (iii) Laboratory Facility Upkeep and Maintenance, and (iv) Safety and Environmental Concerns.

These four topics are listed and described within the text of the chapter along with some notes and clarifications as appropriate. The list of items under each topic is fairly comprehensive but may not be totally inclusive of all those components, which may constitute the sub element.

Contents of the Sub Element

The four topics mentioned earlier are presented in Table 5.1 in the form of questions related to each topic. For each topic, additional detail or "points to consider" are offered along with notes and clarifications as appropriated. The questions, notes, and clarifications are all based on actual inspectional findings from regulatory inspectors or points, which have been found to directly or indirectly impact compliant and efficient real-world laboratory control system operations.

Laboratory Control System Operations in a GMP Environment, First Edition. David M. Bliesner.

Companion website: www.wiley.com/go/Bliesner/LabControl_GMPEnvironment

Table 5.1 Components of the laboratory facilities sub element.

#	Laboratory facilities sub element topic	Notes and clarifications
5.1	***Physical Construction and Layout of the Laboratory***	
5.1.1	Is there adequate space to support all testing and testing related activities?	The GMPs state "C--Buildings and Facilities Sec. 211.42 Design and construction features. (a) Any building or buildings used in the manufacture, processing, packing, or holding of a drug product shall be of suitable size, construction, and location to facilitate cleaning, maintenance, and proper operations." This includes laboratory spaces as well.
5.1.2	Is the laboratory design and layout optimized for the following:	
5.1.2.1	Logical work flow which maximizes efficiency?	Organizing a laboratory along the lines of testing work flow can greatly enhance efficiency and compliance to the GMPs. Unfortunately, many labs do not capitalize on this approach and end up wasting lab space and analysts' time.
5.1.2.2	Minimizing cross-contamination during sample preparation and testing?	Analysts should have sufficient bench space as to not encroach into another analysts' space. A cramped laboratory space is a red flag to regulatory inspectors.
5.1.2.3	Minimizing sample and standard solution mix-ups during preparation and testing?	
5.1.3	Is the laboratory workspace constructed and furnished appropriately with consideration given to the following:	
5.1.3.1	Composition of casework, cabinets, shelving, benchtops, drawers, etc., which may include:	Cost differentials between the materials listed below can be significant. Therefore, an assessment must be made to determine if it is necessary to spend money for a chemically inert bench top or sink (e.g. phenolic resin) or go with a cheaper substitute like plastic laminate.

(Continued)

Table 5.1 (Continued)

#	Laboratory facilities sub element topic	Notes and clarifications
5.1.3.1.1	Stainless steel?	
5.1.3.1.2	Phenolic resin?	
5.1.3.1.3	Black epoxy?	
5.1.3.1.4	Painted steel?	
5.1.3.1.5	Laminated wood?	
5.1.3.1.6	Plastic laminate?	
5.1.3.2	Composition and design of flooring including type of materials used and presence of drains?	Laboratories should be designed to include a drain which is lowest point in the room. This will prevent flooding in the event of a water pipe break or the use of a safety shower.
5.1.3.3	Composition of ceiling tiles?	Acoustical tiles; standard or cleanable.
5.1.3.4	Composition of balance tables?	Marble or solid epoxy.
5.1.3.5	Composition of sinks?	Stainless steel or epoxy.
5.1.3.6	Composition of bookcases?	Wood laminate or painted steel.
5.1.4	Is access to the laboratory restricted and does the security consist of at least one or more of the following:	
5.1.4.1	Security card access?	
5.1.4.2	Biometrics?	Finger print scanner.
5.1.4.3	Lock-and-key?	
5.1.4.4	Mixed means?	Uses a combination of two or more locks.
5.1.4.5	Logbook sign-in and sign-out?	
5.1.5	Has space been allocated in the laboratory for the following activities or items:	
5.1.5.1	Sample and standard preparation?	
5.1.5.2	A balance or weighing room?	If possible, the laboratory should have a separate weighing room where vibrations are minimized and the temperature and humidity are more tightly controlled.
5.1.5.3	Reagent storage?	This includes storing reagents which are not compatible (e.g. acids and bases) in separate areas. Storage areas should be ventilated as appropriate.

(Continued)

Table 5.1 (Continued)

#	Laboratory facilities sub element topic	Notes and clarifications
5.1.5.4	Solvent storage?	Solvent storage areas are often ventilated.
5.1.5.5	Secure sample storage, including:	With the industry currently struggling with data integrity issues, sample chain of custody for all samples (not just controlled substances) has become a requirement.
5.1.5.5.1	Initial receipt?	
5.1.5.5.2	On-test?	
5.1.5.5.3	Pending data review?	
5.1.5.5.4	Hold for investigation?	
5.1.5.5.5	For disposal?	
5.1.5.6	Temporary waste storage?	
5.1.5.7	Prepared solution (volumetric, standardization, etc.) storage?	
5.1.5.8	Controlled substance storage?	For DEA scheduled materials; Schedule I through V.
5.1.5.9	HPLC and GC column storage?	HPLC and GC columns should be stored in a limited access manner to prevent "swap-outs" by analysts at will. Moreover, a detailed log needs to be kept documenting the history of the columns use, including test method, mobile phase, sample types, number of injections, and what solvent the column is stored in.
5.1.5.10	Storage of laboratory expendable supplies (e.g. gloves, filters, vials, etc.)?	
5.1.5.11	Glassware storage?	
5.1.5.12	Refrigerators?	
5.1.5.13	Freezers?	

(Continued)

Table 5.1 (Continued)

#	Laboratory facilities sub element topic	Notes and clarifications
5.1.5.14	Equipment repair parts locker?	In order to speed up the process of equipment repairs, it is sometimes useful to maintain an equipment parts locker on site. The parts locker contains the most common replacement parts, and may also include parts, which may take a long time to receive if ordered at the time equipment failure. Parts lockers can be maintained by laboratory personnel or by the Vendor who performs repairs.
5.1.6	Do analysts have access to office space which includes the following:	
5.1.6.1	Cubicles or offices with sufficient desk space to perform data analysis, report write-up, etc.?	Securable desk drawer space should be made available so analysts can secure notebooks, worksheets, and other GMP documents at night or when they are not actually using them.
5.1.6.2	Networked computers and printers with access to SOPs, test methods, and specifications?	Hard copies, which are properly controlled through the document and change control system, are acceptable as well.
5.1.6.3	Ergonomically fashioned furniture including chairs, desks, or workstations?	
5.1.6.4	Book shelves stocked with appropriate, and current, chemical, safety, and compliance reference literature?	If not accessible in an online format, current copies of pharmacopeia (USP, BP, EP) should be available.
5.1.6.5	Secure filing cabinets to store GMP documentation as appropriate?	Copies of method validation reports, printed specification sheets, etc.
5.1.6.6	Reasonable access, or proximity to, restrooms?	
5.1.6.7	Reasonable access, or proximity to, locker or changing rooms?	If locker or changing rooms exist outside of the manufacturing area.
5.1.6.8	Reasonable access, or proximity to, break rooms or lunch rooms?	
5.1.6.9	Reasonable access, or proximity to, potable water?	Drinking fountains or water coolers.

(Continued)

Table 5.1 (Continued)

#	Laboratory facilities sub element topic	Notes and clarifications
5.2	***Utilities and Services***	
5.2.1	Have the following points been addressed regarding laboratory **water** systems:	
5.2.1.1	Definition of types of water supplied to the laboratory including:	
5.2.1.1.1	Tap water?	Typically, available via laboratory sink taps.
5.2.1.1.2	Purified water?	Typically, available by a laboratory water purification system. See USP definition for purified water and USP <1231> WATER FOR PHARMACEUTICAL PURPOSES for more details.
5.2.1.2	Has the laboratory purified water system been validated?	See USP <1231> WATER FOR PHARMACEUTICAL PURPOSES.
5.2.1.3	Is the quality of laboratory purified water routinely monitored?	For QC laboratories this typically means monitoring, resistivity, filter cartridge life and, in some cases, total organic carbon (TOC).
5.2.1.4	Is preventive maintenance performed on the laboratory purified water system on a periodic, scheduled basis?	Such as filter cartridge replacement.
5.2.1.5	Is purified water used for testing? (e.g. for preparing sample and standard solutions, mobile phases, etc.)?	
5.2.1.6	Is purified water used to wash laboratory glassware?	
5.2.1.7	Is there a procedural contingency to use bottled purified water in the event the in-house laboratory water system becomes unserviceable?	If using bottle reagent water, be cognizant of the fact that microbes will start growing within 24 h of opening the bottle.
5.2.2	Have the following points been addressed regarding laboratory **electrical power** supplies:	
5.2.2.1	Does the laboratory have sufficient access to electrical power (e.g. number of outlets with appropriate voltage and amperage ratings) to power all equipment needed to perform all assigned testing and tasks?	

(Continued)

Table 5.1 (Continued)

#	Laboratory facilities sub element topic	Notes and clarifications
5.2.2.2	Does the laboratory have uninterruptible power supplies (UPSs) for equipment which is critical to testing (e.g. HPLCs, UPLCs, etc.)?	The system should provide uninterruptable power to ensure no disruption in testing due to power "pops" and fluctuations. These systems typically have a limited time in which they can operate when there is a total power failure.
5.2.2.3	Has the uninterruptable power supply system been validated?	
5.2.2.4	Does the laboratory have a backup power system which is validated and periodically tested?	
5.2.3	Have the following points been addressed regarding laboratory **internet and network** access:	
5.2.3.1	Does the laboratory have access to the internet?	
5.2.3.2	Is the laboratory access to the internet restricted and protected (e.g. anti-virus and firewalls) to prevent accidental security breaches?	
5.2.3.3	Do laboratory personnel have access to local and corporate networks?	
5.2.3.4	Is access to local and corporate networks appropriately controlled to prevent unauthorized access and modification to files and folders, which reside on the servers?	
5.2.4	Have the following points been addressed regarding heating, ventilation, and air-conditioning (**HVAC**):	
5.2.4.1	Have temperature monitoring systems and equipment been properly validated?	
5.2.4.2	Have humidity monitoring systems and equipment been properly validated?	
5.2.4.3	Are all controlled temperature/humidity areas monitored to assure proper conditions are maintained?	

(Continued)

Table 5.1 (Continued)

#	Laboratory facilities sub element topic	Notes and clarifications
5.2.4.4	Is temperature and humidity consistently maintained in the monitored areas and in the laboratory in general?	Temperature and humidity fluctuations in older facilities can be a substantial challenge. These fluctuations can dramatically impact testing. For example, the author has observed dimethyl sulfoxide (DMSO) and glacial acetic acid, solidifying during sample preparation in poorly insulated labs during the winter in northern climates.
5.2.4.5	Are laboratory hoods and exhaust arms (e.g. snorkel trunks):	
5.2.4.5.1	Used as the manufacturer intended (e.g. correct sash levels, items not stored in the hood, etc.)?	
5.2.4.5.2	Properly balanced?	
5.2.4.5.3	Properly exhausted?	Hood exhaust should not be near the intake, else the unwanted exhaust fumes will reenter the building.
5.2.4.5.4	Routinely cleaned?	
5.2.4.5.5	Periodically maintained?	
5.2.5	Are the following **gases** used in the laboratory:	
5.2.5.1	Nitrogen?	
5.2.5.2	Argon?	
5.2.5.3	Helium?	
5.2.5.4	Hydrogen?	
5.2.5.5	Compressed air?	
5.2.5.6	If any of the gases listed above are provided through an in-house piping system, have those systems been qualified to ensure purity of the gases?	Compressed air is notorious for containing traces of compressor pump oil, especially if it was not properly installed and qualified.
5.2.5.7	If the gases listed earlier are provided via gas cylinders, are those cylinders secured in a proper fashion and the plumbing checked for leaks when tanks are added or removed?	

(Continued)

Table 5.1 (Continued)

#	Laboratory facilities sub element topic	Notes and clarifications
5.2.6	Have the following points been addressed regarding **piping** systems located in the laboratory:	
5.2.6.1	Are all pipes correctly labeled including the material they carry and the direction of flow?	
5.2.6.2	Have all pipes been cleaned, flushed, passivated, leak tested, and certified for use by the engineering department or contractor/vendor who installed them?	
5.2.7	Have the following points been addressed regarding **disaster planning** for the laboratory:	
5.2.7.1	Does the laboratory have a formal disaster recovery plan?	
5.2.7.2	Has the disaster recovery plan been tested?	
5.2.7.3	Does the disaster recovery plan include contingencies to use backup/alternate laboratory locations within the company?	
5.2.7.4	Does the disaster recovery plan include contingencies to use backup/alternate contract laboratory locations?	
5.3	***Laboratory Facility Upkeep and Maintenance***	
5.3.1	Does laboratory management perform daily walkthroughs of the laboratory?	
5.3.2	Does quality assurance (QA) perform periodic general GMP compliance practice audits of the laboratory?	
5.3.3	Does the laboratory have a "benchtop patrol program" where designated personnel, look for on a weekly or monthly basis, the presence of dust, dirty glassware, expired solutions, unlabeled or expired squirt bottles, and general laboratory compliance practices?	Squirt bottles containing water must be dumped and refilled every 24 hours due to microbial growth.

(Continued)

Table 5.1 (Continued)

#	Laboratory facilities sub element topic	Notes and clarifications
5.3.4	Are the following systems or pieces of equipment included in a scheduled preventive maintenance program:	
5.3.4.1	Laboratory water system filter cartridge changes (e.g. ion exchange resins, activated carbon filters, etc.)?	
5.3.4.2	HVAC filter changes?	
5.3.4.3	Glassware washer maintenance?	
5.3.4.4	Hood function checks (e.g. periodic smoke testing)?	
5.4	***Safety and Environmental Concerns***	Although not a GMP requirement, safety is a good indicator of a laboratory's state of GMP compliance. If a lab doesn't worry about safety, they invariable don't care about GMP compliance.
5.4.1	Is there a laboratory safety and environmental SOP?	
5.4.2	Are laboratory personnel trained in laboratory safety?	
5.4.3	Do laboratory personnel have immediate access to current material safety data sheets (MSDSs)?	
5.4.4	Are eyewash stations readily available, and flushed and checked for functionality on a regular basis?	Eyewash stations must be regularly flushed to prevent microbial and parasitic growth.
5.4.5	Are safety showers easily accessible and checked for functionality on a regular basis?	
5.4.6	Are chemical spill kits available, accessible, and inventoried on a periodic basis?	
5.4.7	Are fire blankets easily accessed?	
5.4.8	Are hazardous materials identified, segregated, and stored in an appropriate fashion?	
5.4.9	Is there a program that describes handling and disposal of liquid chemical waste in compliance with Local, State, and Federal regulations?	

(Continued)

Table 5.1 (Continued)

#	Laboratory facilities sub element topic	Notes and clarifications
5.4.10	Is there a program that describes handling and disposal of solid chemical waste in compliance with Local, State, and Federal regulations?	
5.4.11	Are HPLC vials disposed of in compliance with Local, State, and Federal regulations?	
5.4.12	Is there a program that describes handling and disposal of non-chemical waste in compliance with Local, State, and Federal regulations?	
5.4.13	Are laboratory personnel trained to handle hazardous materials and laboratory waste?	
5.4.14	Is there a procedure in place which addresses the purchase, receipt, cleaning, and storage of laboratory coats?	

Tools and Templates

The following is provided in electronic format as an Example Tool and Template, which can be found in the Chapter 5 Appendix (www.wiley.com/go/Bliesner/LabControl_GMPEnvironment).

- Table 5.1 "Components of the laboratory facilities sub element" in checklist format for use in evaluating, upgrading, or establishing your own laboratory facilities system.

References

1 Thermo Fisher Scientific Inc. (2018). *Fisherbrand Lab Furnishings Catalog*. Thermo Fisher Scientific Inc, https://online.uspnf.com/uspnf.

2 USP Purified Water (2015). GCCA2015 General Chapters-Chemical Analysis.

3 USP <1231> WATER FOR PHARMACEUTICAL PURPOSES.

4 Scientific Equipment & Furniture Association (2013). SEFA 8-P-2013 Recommended Practices for Polypropylene Laboratory Grade Furniture, Casework, Shelving and Tables. Garden City, NY.
5 Drug Scheduling. United States Drug Enforcement Administration. www.DEA.gov.
6 Laboratory Countertops & Bench Tops. LFFH, Inc. https://labfurnitureandfumehoods.com/laboratory-epoxy-resin-countertops/

6

Components of the Method Validation and Method Transfer Sub Element (MV)

Description of the Method Validation and Method Transfer Sub Element

Method validation is the process of demonstrating that analytical test procedures are suitable for their intended use. Method validations have traditionally been executed by research and development (R&D) organizations following comprehensive method developments studies. Upon conclusion of a successful method validation, the validated test methods are transferred to quality control (QC) testing labs where they become the backbone of QC operations.

The process of validating an analytical method begins with creation and approval of a method validation protocol. This protocol is reviewed and approved by Quality Assurance (QA) and delineates precisely how the validation exercise will proceed, including pre-approved acceptance criteria. Following the planned and systematic collection of validation data, test methods are finalized, formalized, and transferred, via method transfer protocols, from R&D (the originating laboratory) to QC laboratory (the receiving laboratory).

Method Validation and Method Transfer include at least five individual topics. A laboratory, which is in compliance with Current Good Manufacturing Practices (CGMPs), should at least address each of these topics. These topics include (i) Regulatory Requirements and Industry Guidance Related to Analytical Method Validation, (ii) Analytical Method Development, (iii) Analytical Method Validation, (iv) Verification of Compendial Procedures, and (v) Analytical Method Transfer.

It should be noted that this chapter is not written to be an all-inclusive description of how to develop, validate, and transfer an analytical test method. Entire texts have been written, which describe in great detail this process [4]. Instead this chapter is intended to:

- Familiarize QC and QA personnel as to the regulatory requirements for developing, validating, and transferring an analytical test method

Laboratory Control System Operations in a GMP Environment, First Edition. David M. Bliesner.

Companion website: www.wiley.com/go/Bliesner/LabControl_GMPEnvironment

- Educate QC and QA personnel so that they can be actively and intelligently involved in the method development process and subsequent validation and transfer exercises
- Encourage early involvement of QC personnel in the method development and validation process so that the validated method can be seamlessly transferred to the QC laboratory

On a final note, this chapter does not address the development and validation of methods used to test cleaning samples received from the manufacturing department following their post-production cleaning. These studies are also traditionally executed by R&D organizations with subsequent transfer to the QC testing laboratory who analyzes cleaning samples obtained during cleaning of manufacturing equipment.

Contents of the Sub Element

The five topics are previously listed and described within the text in Table 6.1, along with some notes and clarifications as appropriate.

Table 6.1 Components of the method validation and Method Transfer sub element.

#	Method validation and method transfer sub element topic	Notes and clarifications
6.1	***Regulatory Requirements and Industry Guidance Related to Analytical Method Validation***	
6.1.1	Have the following US regulatory requirements been reviewed and considered during method development and validation exercises:	
6.1.1.1	21 CFR 211.165 Testing and release for distribution "(e) The accuracy, sensitivity, specificity, and reproducibility of test methods employed by the firm shall be established and documented. Such validation and documentation may be accomplished in accordance with 211.194(a)(2)"?	This is extracted from the United States Code of Federal Regulations (CFR) PART 211 CURRENT GOOD MANUFACTURING PRACTICE FOR FINISHED PHARMACEUTICALS.

Table 6.1 (Continued)

#	Method validation and method transfer sub element topic	Notes and clarifications
6.1.1.2	21 CFR 211.194 Laboratory records (a)(2) "A statement of each method used in the testing of the sample. The statement shall indicate the location of data that establish that the methods used in the testing of the sample meet proper standards of accuracy and reliability as applied to the product tested. (If the method employed is in the current revision of the United States Pharmacopeia, National Formulary, AOAC International, Book of Methods, or in other recognized standard references, or is detailed in an approved new drug application and the referenced method is not modified, a statement indicating the method and reference will suffice). The suitability of all testing methods used shall be verified under actual conditions of use"?	This is also extracted from United States Code of Federal Regulations (CFR) PART 211 CURRENT GOOD MANUFACTURING PRACTICE FOR FINISHED PHARMACEUTICALS.
6.1.2	Have the following European regulatory requirements been reviewed and considered during method development and validation exercises:	EudraLex – Volume 4 – Good Manufacturing Practice (GMP) guidelines. Volume 4 of "The rules governing medicinal products in the European Union" contains guidance for the interpretation of the principles and guidelines of good manufacturing practices for medicinal products for human and veterinary use laid down in Commission Directives 91/356/EEC, as amended by Directive 2003/94/EC, and 91/412/EEC respectively.
6.1.2.1	EudraLex – Volume 4 – EU Guidelines for Good Manufacturing Practice for Medicinal Products for Human and Veterinary Use, Part 1, Chapter 6: Quality Control?	

(Continued)

Table 6.1 (Continued)

#	Method validation and method transfer sub element topic	Notes and clarifications
6.1.2.2	Section labeled "Testing" 6.15 "Testing methods should be validated. A laboratory that is using a testing method and which did not perform the original validation, should verify the appropriateness of the testing method. All testing operations described in the marketing authorisation or technical dossier should be carried out according to the approved methods"?	
6.1.3	Have the following industry guidances been reviewed and considered during method development and validation exercises:	
6.1.3.1	U.S. FDA guidance titled "Analytical Procedures and Methods Validation for Drugs and Biologics" July, 2015?	
6.1.3.2	ICH HARMONISED TRIPARTITE GUIDELINE VALIDATION OF ANALYTICAL PROCEDURES: TEXT AND METHODOLOGY Q2(R1) November 2005?	
6.1.3.3	World Health Organization (WHO), GUIDELINES ON VALIDATION – APPENDIX 4 ANALYTICAL METHOD VALIDATION, September 2018?	
6.1.3.4	USP-NF <1224>TRANSFER OF ANALYTICAL PROCEDURES?	
6.1.3.5	USP-NF <1225> VALIDATION OF COMPENDIAL PROCEDURES?	
6.1.3.6	USP-NF <1226> VERIFICATION OF COMPENDIAL PROCEDURES?	
6.1.3.7	USP-NF <621> CHROMATOGRAPHY?	

Table 6.1 (Continued)

#	Method validation and method transfer sub element topic	Notes and clarifications
6.2	***Analytical Method Development***	Method development and method validation are two distinct and different endeavors but are closely linked. Method development is the process by which methods are developed or modified and evaluated for suitability of use as test methods. The resulting method then serves as a precursor for method validation. Method validation is the collection of documented evidence, which demonstrates that a method is suitable for its intended use. Method validation is performed at the completion of method development. It should be noted that ICH Q14 "Analytical Procedure Development Guideline," is a new guideline that was endorsed by the ICH Assembly in June 2018. When it became effective, it will require R&D organizations involved in method development to operate in compliance with the GMPs.
6.2.1	Are the following points taken into consideration with respect to method development:	
6.2.1.1	An assessment of user requirements including:	Users in this case are QC analysts tasked with testing raw materials, active pharmaceutical ingredients (APIs), finished drug product, stability samples, etc. in the receiving laboratory.
6.2.1.1.1	Availability of analytical equipment at users' location?	
6.2.1.1.2	Availability of supplies such as reagents for making mobile phases, HPLC columns, reference standards, etc.?	
6.2.1.1.3	Existence of language barriers?	Final methods must be published in the language commonly used at the site to which they will be transferred.

(Continued)

Table 6.1 (Continued)

#	Method validation and method transfer sub element topic	Notes and clarifications
6.2.1.1.4	Existence of expertise barriers?	Are there analysts at the site who have the technical expertise to execute routine QC release testing using the final transferred method?
6.2.2	Does method development address the following:	
6.2.2.1	What characteristics of the API or drug product are to be evaluated?	Characteristics may include tests for assay (strength), impurities, dissolution, etc.
6.2.2.2	What acceptance criteria are appropriate for specific characteristics?	
6.2.2.3	An assessment of the best analytical technique to be used for testing?	
6.2.2.4	An assessment of the most appropriate analytical instrumentation to be used for testing?	
6.2.2.5	An evaluation of existing, validated methods and validation packages for potential starting points for the current method validation efforts?	
6.2.2.6	A determination of the USP method category under which the method will be evaluated, including:	
6.2.2.6.1	USP category I?	Validation of analytical methods for assay.
6.2.2.6.2	USP category II?	Validation of analytical methods for impurities and degradants.
6.2.2.6.3	USP category III?	Validation of analytical methods for dissolution.
6.2.2.6.4	USP category IV?	Validation of analytical methods for identification.
6.2.2.7	An evaluation of existing skill sets and expertise of laboratory personnel who will support the method development, validation and method transfer efforts?	This evaluation refers to the QC personnel at the receiving site who are involved in the method transfer.
6.2.2.8	Evaluation of API manufacturer technical reports, documents, and previous validation efforts?	

Table 6.1 (Continued)

#	Method validation and method transfer sub element topic	Notes and clarifications
6.2.2.9	Review of existing scientific literature related to the analyses of the API or finished drug product?	
6.2.2.10	Review of existing scientific literature related to the physical-chemical properties of the API or finished drug product?	
6.2.2.11	Review of existing scientific literature related to the physical–chemical interactions of the API with excipients?	Excipient compatibility studies.
6.2.2.12	Execution of a systematic approach to method robustness as part of the method development exercise?	According to FDA "…robustness of methods should be evaluated because this characteristic can help you decide which method you will submit for approval." [5]
6.2.3	Are method development reports created at the conclusion of method development efforts?	
6.2.4	Are development data which support validation of methods included in the method validation report?	Development data should be submitted within the method validation if they support the validation of the method.
6.3	***Analytical Method Validation***	According to FDA "Data must be available to establish that the analytical procedures used in testing meet proper standards of accuracy, sensitivity, specificity, and reproducibility and are suitable for their intended purpose."
6.3.1	Is there an SOP which describes how methods are to be validated?	
6.3.2	Are the requirements in the SOP consistent with FDA, ICH, and USP guidelines?	
6.3.3	Are method validation protocols created, approved, and executed for each validation?	Protocols must be reviewed and approved by QA.
6.3.4	Are all of the appropriate elements of validation included in the method validation protocol such as:	These are also referred to as analytical performance characteristics.

(Continued)

Table 6.1 (Continued)

#	Method validation and method transfer sub element topic	Notes and clarifications
6.3.4.1	Accuracy?	Please see the glossary at the end of this chapter for definitions of all the analytical performance characteristics.
6.3.4.2	Precision (Repeatability, intermediate precision, and reproducibility)?	
6.3.4.3	Selectivity/specificity?	Selective = separate and show every component in the sample Specific = measure only the desired component without interference from other species which might be present; separation is not necessarily required.
6.3.4.4	Robustness?	
6.3.4.5	Linearity?	
6.3.4.6	Range?	
6.3.4.7	Limit of detection?	
6.3.4.8	Limit of quantitation?	
6.3.4.9	System suitability?	
6.3.4.10	Sample solution stability?	
6.3.4.11	Completeness of extraction (from dosage forms)?	
6.3.4.12	Filter retention studies?	
6.3.4.13	Forced degradation studies?	
6.3.4.14	Analysis of aged samples?	This includes development finished dosage forms which have been aged or subject to more aggressive conditions than what is expected to occur in the market place.
6.3.5	Are method validation reports published following validation of an analytical method?	Reports should be reviewed and approved by QA.
6.3.6	Can results presented in the method validation reports be tracked back to raw data?	Raw data may be in paper, electronic, or hybrid format.

Table 6.1 (Continued)

#	Method validation and method transfer sub element topic	Notes and clarifications
6.3.7	Does the validation SOP address the content of the final analytical procedure (e.g. Official Test Method)?	This includes sections for scope, apparatus/equipment, operating parameters, reagents, standards, sample preparation, standard solution preparation, steps in the procedure, system suitability, calculations, and reporting of results.
6.3.8	Does the method validation SOP address lifecycle management of analytical procedures?	This ensures the procedure remains fit for its intended purpose over time. This may include trend analysis of method performance, review, and risk assessments for changes in the product manufacturing scheme, changes in availability of supplies used during analysis (e.g. HPLC columns become no longer available, etc.), and other changes.
6.3.9	Does the method validation SOP address revalidation of methods?	See FDA guidance for further explanation. [5]
6.4	***Verification of Compendial Procedures***	See USP <1226> VERIFICATION OF COMPENDIAL PROCEDURES for additional detail and instruction.
6.4.1	Is there an SOP or section within an SOP, which discusses the requirements for the verification of Compendial methods?	
6.4.2	Are verification requirements based on an assessment of the complexity of both the procedure and the material to which the procedure is applied?	
6.4.3	Does the procedure require that compendial methods have been shown to work under the actual conditions of use?	This includes analysis of specified drug substance and/or drug product.
6.4.4	Does the verification procedure require a pre-approved verification protocol, with pre-established acceptance criteria?	
6.4.5	Is a verification report generated at the completion method verification process?	

(Continued)

Table 6.1 (Continued)

#	Method validation and method transfer sub element topic	Notes and clarifications
<u>***6.5***</u>	<u>***Analytical Method Transfer***</u>	See FDA guidance titled "Analytical Procedures and Methods Validation for Drugs and Biologics" and USP <1226> VERIFICATION OF COMPENDIAL PROCEDURES for additional detail and guidance.
6.5.1	Are there site policies and procedures governing transfer of methods, and do these documents include defining acceptance criteria?	
6.5.2	Are the responsibilities for transfer clearly defined (e.g. receiving laboratory or originating laboratory)?	
6.5.3	Is there a requirement to form an analytical method transfer team or similar organization to transfer methods?	
6.5.4	Is a formal receiving laboratory readiness assessment performed to include evaluation of equipment, facilities, and personnel?	
6.5.5	Is the receiving laboratory given the opportunity to review the analytical methodology prior to the transfer process?	
6.5.6	Is method transfer protocol, with pre-determined acceptance criteria used to formally transfer the methodology?	Method transfer can take the form of (i) Comparative sample testing, (ii) Co-validation of the method or (iii) Re-validation of the method at the receiving laboratory.
6.5.7	Are personnel in the receiving laboratory given the opportunity to use the method under the supervision of the originating personnel subject matter experts prior to formal transfer?	
6.5.8	Are deviations from the transfer protocol and/or failures to meet acceptance criteria documented, investigated, and reported?	

Table 6.1 (Continued)

#	Method validation and method transfer sub element topic	Notes and clarifications
6.5.9	Is a statistical comparison between acceptance criteria and final results performed to ensure the method transfer was successful?	The data analysis (e.g. Linear regression, two-one-sided *t*-tests, etc.) should be assured either by using appropriately validated software or independent verification for correctness.
6.5.10	Is a reviewed and approved method transfer report generated upon successful execution of the transfer protocol?	The report should contain the following sections: (i) Results and discussions, (ii) Deviations and investigations, (iii) Conclusion, and (iv) References.

Tools and Templates

The following are provided in electronic format as Example Tools and Templates, which can be found in the Chapter 6 Appendix (www.wiley.com/go/Bliesner/LabControl_GMPEnvironment):

- Table 6.1 "Components of the Method Validation and Method Transfer Sub Element" in checklist format for use in evaluating, upgrading or establishing your own method validation and method transfer approach
- Example-Template method validation report

Glossary

The following terms are commonly encountered while working in a CGMP Laboratory during the execution of a methods validation:

483 (Form 483) The designation of the FDA form, which is used to capture and report observations of CGMP deficiencies related to FDA on-site audits. These observations are compiled and become part of the Facilities Inspection Report. Form 483 observations become part of the public record.

Acceptance Criteria Numerical limits, ranges, or other suitable measures used to determine the acceptability of the results of analytical procedures.

Accuracy	Expresses the closeness of agreement between the value found and the value that is accepted as either a conventional true value or an accepted reference value. It may often be expressed as the recovery by the assay of known, added amounts of analyte.
Action Level/Alert Level	Alert level is used to identify the point at which a parameter has drifted toward the extreme of the specified operating range. Action level is when the parameter has drifted outside of the specified operating range. Alert and action levels must be tighter than registration specifications. Alerts are reported to management and evaluated. If an action level is reached, it is reported to management, investigated and a corrective action initiated.
Active Pharmaceutical Ingredient (API)	Also known as drug substance. Any component that is intended to furnish pharmacological activity or other direct effect in the diagnosis, cure, mitigation, treatment, or prevention of disease, or to affect the structure of any function of the body of man or other animals.
Analytical Performance Characteristics	A term used by the USP; Analytical performance characteristics refers to those characteristics of an analytical method, which define its performance as an analytical technique. These performance characteristics include accuracy, precision, specificity, detection limit, quantitation limit, linearity, and range. These characteristics need to be considered when validating anyone of the USP method categories.
API	See Active Pharmaceutical Ingredient (API).
Atypical Result	Results generated on a test article that are within specifications but are inconsistent with previous data, established trends, or other results for the same sample on test.
Audit Summary Report (ASR)	The final output of the laboratory audit. A coherent and organized presentation of findings and suggestions for corrective and preventive actions.

Batch A specific quantity of a drug or other material that is intended to have a uniform character and quality within specified limits and is produced according to a single manufacturing order during the same one cycle of manufacture.

Batch Record A record prepared for each batch of drug product or API produced that includes complete information relating to the production and control of each batch.

Blank A sample or standard of a particular matrix or composition without analyte.

Buffer (Buffering Agent) A substance or mixture of substances (such as phosphate salts) that in solution tend to stabilize the hydrogen ion concentration by neutralizing within limits both acids and bases, so the solution resists changes to pH.

Calibration The demonstration that a particular instrument or device produces results within specified limits by comparison with those produced by a reference or traceable standard over an appropriate range of measurements.

Calibration Curve A calibration curve is a plot of standard solution concentration on the x-axis versus instrument response on the y-axis. In chromatographic analyses calibration curves are generated by analyzing standard analyte solutions of known concentration and measuring the resulting chromatographic peak area. The resulting plot is then used to determine the concentration of unknown sample solutions containing the same analyte. This is done by measuring the unknown peak area (y) and using the equation for the line to solve for the concentration of the unknown (x). Although referred to as a "curve," it is usually a linear plot with a well-defined slope and y-intercept.

CAPA (Corrective and Preventive Actions) The steps taken to correct and prevent deficiencies uncovered during a laboratory audit.

Capacity Factor *k′* A dimensionless quantity used to describe the retention of a compound. It is calculated by the following formula: $(t_r - t_0)/t_0$, where t_r is the measured retention time of the component of interest, and t0 is the retention time of an un-retained component. t0 is most measured at the first disturbance of the baseline in HPLC analyses. t_r is measured at the peak apex. k' is a normalized value for retention. Values range between 2 and 10 for acceptable chromatography.

CAPP (Corrective Action Project Plan) The project plan generated to insure successful implementation of corrective and preventive actions.

CBE 30 (Change Being Effected, 30 Days) Supplemental changes to applications which do not require prior approval by FDA. These changes may be implemented within thirty days following submission to FDA if the agency has no comments.

CFR (Code of Federal Regulations) The Code of Federal Regulations (CFR) is the codification of the general and permanent rules published in the Federal Register by the executive departments and agencies of the Federal Government. It is divided into 50 titles that represent broad areas subject to Federal regulation.

CGMPs (Current Good Manufacturing Practices) 21 Code of Federal Regulations Parts 210 and 211. Federal regulations, which describe the minimum current good manufacturing practices for preparation of drug products for administration to human and animals. They include methods to be used in and the facilities or controls to be used for the manufacturing, processing, packing, or holding of a drug to assure that such drug meets the requirements of the act as to safety, and has the identity and strength and meets the quality and purity characteristics that it purports or is represented to possess.

Change Control Procedure A procedure describing measures to be taken for the purpose of controlling and maintaining an audit trail when changes are made to any part of a system (e.g. standard operating procedure, test method, or specification).

Check Standard	A second preparation of the working standard, which is analyzed as part of the system suitability run. The check standard is prepared at the same concentration as the working standard. Prior to continuing the chromatographic run the ratios of the response factors (response factor = area/concentration) for the working standard and the check standard is calculated. RF check standard/RF working standard should normally be within ±2.0%. This provides assurance that the working standard was prepared correctly.
Compendial Tests Methods	Test methods that appear in official compendia such as the United States Pharmacopoeia (USP/NF).
Complaint	Any verbal, written, or electronic report that alleges deficiencies related to the identity, strength, quality, purity, or effectiveness of a product after it has been released for distribution.
Component	Any ingredient intended for use in the manufacture of a drug product including those that may not appear in such drug product and primary packaging components.
Compounding	Bringing together into homogenous mix of active ingredients, excipients, and, as applicable, solvent components.
Consent Decree	A voluntary legal agreement a drug firm enters into with the FDA for the expressed purpose of correcting deficiencies related to CGMPs.
CRO (Contract Research Organization)	A contract manufacturing or analytical testing company.
Degradation Product	A molecule resulting from a chemical change in the drug molecule brought about over time and/or the action of light, temperature, pH, water, etc. or by reaction with and excipient and/or the immediate container/closure system.
Denaturation	A condition in which a protein unfolds or its polypeptide chains are disordered, rendering the molecule less soluble and usually nonfunctional. The amount of organic modifier in reverse phase HPLC, if high enough, can denature proteins.

Detection Limit	The detection limit (DL or limit of detection LOD) of an individual procedure is the lowest amount of analyte in a sample that can be detected but not necessarily quantitated as an exact value. The LOD is a parameter of limit tests (tests that only determine if the analyte concentration is above or below a specification limit). In analytical procedures that exhibit baseline noise, the LOD can be based on a signal to noise ratio (3 : 1), which is usually expressed as the concentration (e.g. percentage, parts per billion) of analyte in the sample.
Development Report	A report that summarizes the major stages of drug product or API development from early stages through large scale manufacturing.
Documentation	Any combination of text, graphics, data, audio, or video information that can be used to clearly and completely recreate an activity, event, or process. Documentation includes lifecycle documents as well as records.
Document Control System	A system for managing preparation, review, approval, issuance, distribution, revision, retention, archival, obsolescence, and destruction of lifecycle documents.
Drug Product	The combination of API and excipients that are processed into a dosage form and marketed to the public. Common examples include tablets, capsules, oral solutions, etc. Also referred to as finished product or dosage form.
Drug Substance	See Active Pharmaceutical Ingredient (API).
Effective Date	Date by which the approved standard or procedure shall be implemented and in use. All required training must be completed prior to this date.
Excipient(s)	An excipient is a raw material that may perform a variety of roles in a drug product (e.g. tablet press lubricant, filler, diluent, disintegration accelerator, colorant, etc.). However, unlike the API, which is pharmacologically active, the excipient has no intrinsic pharmacological activity.

Expiration Date The date placed on the container/labels of a drug product or API designating the time during which a batch of product is expected to remain within the approved shelf life specification if stored under defined conditions, and after which it must not be used.

Extraction Efficiency A measure of the effectiveness of extraction of the drug substance from the sample matrix. Studies are conducted during methods validation to determine that the sample preparation scheme is sufficient to ensure complete extraction without being unnecessarily excessive. This is normally investigated by varying the shaking or sonication times (and/or temperature) as appropriate.

FDA (Food and Drug Administration) The FDA is responsible for protecting the public health by assuring the safety, efficacy, and security of human and veterinary drugs, biological products, medical devices, our nation's food supply, cosmetics, and products that emit radiation. FDA is also responsible for advancing the public health by helping to speed innovations that make medicines and foods more effective, safer, and more affordable; and helping the public get the accurate, science-based information they need to use medicines and foods to improve their health.

Filter Study A comparison of filter to unfiltered solutions in a methods validation to determine whether the filter being used retains any active compounds or contributes unknown compounds to the analysis.

Forced Degradation Studies Studies undertaken to degrade the sample (e.g. drug product or API) deliberately. These studies, which may be undertaken during method development and/or validation, are used to evaluate an analytical method's ability to measure an active ingredient and its degradation products without interference and are an integral part of the validation of a method as being specific and stability indicating.

Formulation	The recipe which describes the quantity and identity of API and excipients which make up a drug product. For example, a 100 mg tablet of Advil® may actually weigh 165 mg total. The formulation may include: 100 mg of ibuprofen (the active ingredient), 50 mg of starch (filler), 5 mg of talc (lubricant for the tablet press), and 10 mg of iron oxide (colorant).
Good Documentation Practices	The handling of written or pictorial information describing, defining, specifying, and/or reporting of certifying activities, requirements, procedures, or results in such a way as to ensure data integrity.
ICH	The International Council for Harmonisation of Technical Requirements for Pharmaceuticals for Human Use (ICH), formerly called the Tripartite International Conference on Harmonization, is an international organization formed to establish uniform guidance within the pharmaceutical industry. For drug development and manufacture the ICH issues guidance such as ICH Q2A *Text on Validation of Analytical Procedures* (March 1995), which is designed to instill uniformity within the industry with respect to various drug development and manufacturing issues. ICH is composed of industry experts who work jointly and with FDA to develop these guidance documents.
Identification Test	An analytical method that is capable of determining the presence of the analyte and can discriminate between closely related compounds.
In-Process Control/Test	Checks performed during production in order to monitor and, if appropriate, to adjust the process and/or to ensure that the product meets its specifications. The control of the environment or equipment may also be regarded as part of in-process control.
Intermediate	Any substance, whether isolated or not, that is produced by chemical, physical, or biological action at some stage in the production of a drug product or API and that is subsequently used at another stage of production.

Internal Audit	A systematic examination conducted by an internal organizational unit to determine whether quality activities and related results comply with policies, standards, and requirements and whether practices have been implemented effectively and are suitable to achieve objectives and ensure compliance with regulatory requirements.
IQ/OQ/PQ (Installation, Operational, and Performance Qualification)	The process by which laboratory equipment is properly installed, and determined to be operating within specifications for its intended use. IQ/OQ/PQ is executed via protocol with predetermined acceptance criteria.
Label Claim	Theoretical strength of the product as given on the marketed product label.
Laboratory Qualification	A process where a laboratory that has demonstrated that it has the systems in place necessary to properly perform the tests being conducted.
LAF (Laboratory Audit Form)	The primary data capture instrument used when conducting a laboratory Audit.
LIMS (Laboratory Information Management System)	Any computer system that is used to collect, compile, organize, and report laboratory data. LIMS systems may also be used to calculate results as well.
Linearity	Evaluates the analytical procedure's ability (within a given range) to obtain a response that is directly proportional to the concentration (amount) of analyte in the sample. If the method is linear, the test results are directly, or by well-defined mathematical transformation, proportional to the concentration of analyte in samples within a given range. Linearity is usually expressed as the confidence limit around the slope of the regression line. The line is calculated according to an established mathematical relationship from the test response obtained by the analysis of samples with varying concentrations of analyte. Linearity may be established for all active substances, preservatives, and expected impurities. Evaluation is performed on standards. Note that this is different than *Range* (sometimes referred to as *Linearity of Method*), which is evaluated using samples and must encompass the specification range of the component assayed in the drug product.

LIR (Laboratory Investigation Report)	An investigation of any laboratory results or observation, which does not meet acceptance criteria or falls outside the expected operational parameters. Similar to a Out-of-specification (OOS) investigation, but does not have to do with meeting a release specification.
Lot	A batch or any portion of a batch having uniform character and quality within specified limits or, in the case of a drug product is a continuous process, it is a specific identified amount produced in a unit of time or quantity in a manner that assures it having uniform character and quality within specified limits.
Master Production Record	A document that specifies complete manufacturing and control instructions, sampling, and testing procedures, specifications, special notations, and precautions to be followed in production of the product and assures uniformity from batch to batch by specifying the batch size, components and quantities, and theoretical yields.
Matrix (Sample Matrix)	The components and physical form with which the analyte of interest is intimately associated. In the case of drug product, the matrix is the combination of excipients in which the active ingredient is diluted and formed within. For example, the matrix of a transdermal patch is an adhesive, in which the drug substance is dissolved, fixed to a plastic backing covered by a release liner. The chemical composition and physical structure of the matrix can have a substantial effect on sample preparation and extraction of the active moiety.
Maximum Allowable Residue	The maximum allowable residue is used to calculate the acceptance criteria for cleaning verification methods. Residue limits should be practical, achievable, verifiable, and based on the most deleterious residue. Limits may be established based on the minimum known pharmacological, toxicological, or physiological activity of the API or its most deleterious component.

Method Development	The process by which methods are developed and evaluated for suitability of use as test methods and as precursor to validation.
Method Qualification	Preliminary method validation conducted during Phase 1, 2, and 3 to support the drug development process and the associated release of clinical trial material.
Method Transfer	The transfer method including demonstration of equivalence according to predetermined criteria between the receiving and transferring sites.
Methods Validation	A series of systematic laboratory studies where the performance characteristics of analytical procedures are established to meet the requirements for intended analytical applications. FDA states in their guidance document that "Methods validation is the process of demonstrating that analytical procedures are suitable for their intended use. The methods validation process for analytical procedures begins with the planned and systematic collection by the applicant of the validation data to support analytical procedures."
Method Verification	The process by which Compendial methods are determined to be suitable for analysis of a given test article by a given laboratory. As cited in 21 CFR Part 211.194(a) (2), method verification is used for USP-NF Compendial methods or those compendia (and other recognized references such as the AOAC Book of Methods) that have documented evidence that the methods have been validated. At a minimum this involves sample solution stability, specificity, and intermediate precision. Also see USP-NF <1226> VERIFICATION OF COMPENDIAL PROCEDURES.
OOS Investigation	The systematic and planned search for the root cause that generated an Out-of-Specification Result. OOS investigation should include formal reporting and a description of corrective and preventative actions taken.

OOS Result (Out-of-Specification Result) An examination, measurement, or test outcome that does not comply with the specification or predetermined acceptance criteria.

PAI (Pre-Approval Inspection) An inspection by FDA to confirm the CGMP compliance of a drug manufacturing facility. This takes place prior to the FDA's market approval of the drug for sale.

Percent RSD (% RSD) Percent relative standard deviation. A common expression and measure of the relative precision of an analytical method for a given set of measurements. % RSD is calculated by dividing the standard deviation for a series of measurements by the mean of the same sets of measurements and multiplying by 100. % RSD = $(\sigma{\cdot}N^{-1}/\text{mean}) \times 100$. Large % RSDs for a series of measurements indicated significant scatter and lack of precision in the technique.

Personnel Qualification The combination of education, training, and/or experience that enables an individual to perform assigned tasks.

Placebo A formulation containing all ingredients of a drug product except the active ingredient for which the method is being developed.

Precision Expresses the closeness of agreement (degree of scatter) between a series of measurements obtained from multiple aliquots of a homogenous sample under the prescribed conditions. The precision of an analytical procedure is usually expressed as the percent relative standard deviation. According to FDA and ICH, precision may be considered at three levels, namely:

Repeatability: Refers to the use of the analytical procedure within a laboratory by a single analyst, on a single instrument, under the same operation conditions, over a short interval of time. This is sometimes referred to as method precision.

Intermediate Precision. Refers to variations within a laboratory as with different days, with different instruments, by different analysts, and so forth. Formally known has ruggedness.

	Reproducibility. The measure of the capacity of the method to remain unaffected by a variety of conditions such as different laboratories, analysts, instruments, reagent lots, elapsed assay times, days, etc. More succinctly, the use of the method in different laboratories. Reproducibility is not part of the expected methods validation process. It is addressed during technical transfer of the method to different sites.
Process Impurity	Any component of the drug product resulting from the manufacturing process that is not the chemical entity defined as the drug substance or an excipient in the drug product.
Prospective Validation	Validation conducted prior to the distribution of either a new product, or product made under a revised manufacturing process, where the revisions may affect the product characteristics.
Protocol	An approved documented experimental design that, when executed, will demonstrate the ability of the subject method to perform as intended. Formal methods validation studies require a protocol. The protocol must have pre-assigned and approved acceptance criteria for each stage of the validation. The protocol may be a stand-alone document or may reference the methods validation SOP for specific details.
Q	The amount of dissolved active ingredient specified in the monograph, expressed as a percentage of the labeled content of the dosage form, obtained during dissolution testing. For example, a tablet may have a specification stating that 85% of the active must be dissolved in 30 minutes of dissolution testing using USP apparatus II (paddles). In this case $Q = 85\%$. Also, a nefarious character in the *Star Trek: The Next Generation* TV series.
QA	See Quality Assurance Unit.
QC (Quality Control)	The unit responsible for performing testing API and drug product. Often referred to as the QC laboratory.

Qualification Action of proving and documenting that equipment or ancillary systems are properly installed, work correctly, and actually lead to the expected results. Qualification is part of validation, but the individual steps alone do not constitute process validation.

Quality Assurance Unit The Quality Assurance Unit serves the role of the Quality Control Unit as defined in 21 CFR 211.11. In this document, only the compliance function of the Quality Unit is addressed, not the testing functions. In the past there has been some confusion with respect to Quality Assurance and Quality Control. CGMPs now generally recognize QA = compliance, QC = Testing.

Quality System A group of interrelated activities representing an integrated approach to philosophy and practices of manufacturing APIs and drug products to assure safety, identity, strength, purity, and quality. See ICH Q10, PHARMACEUTICAL QUALITY SYSTEM.

Quality Unit An organizational unit independent of production, which ensures that the manufacture, testing, storage, and distribution of drug products, active pharmaceutical ingredients, and components are performed in compliance with regulatory requirements and conformance to company policies and industry practices. Also referred to as Quality Assurance or QA.

Quantitation Limit The quantitation limit (QL or limit of quantitation LOQ) of an individual analytical procedure is the lowest amount of analyte in a sample that can be quantitatively determined with suitable precision and accuracy. The quantitation limit is a parameter of quantitative assays for low concentrations of compounds in sample matrices, and is used particularly for the determination of impurities and/or degradation products. It is usually expressed as the concentration (e.g. percentage, parts per million, etc.) of analyte in the sample. For analytical procedures that exhibit baseline noise the LOQ is generally estimated from a determination of signal-to-noise ratio (10 : 1) and may be confirmed by experiments.

Quarantine — The status of materials isolated physically or by other effective means to preclude their use pending a decision on their subsequent disposition.

Range — The interval between the upper and lower concentrations (amounts) of analyte in the sample (including these concentrations) for which it has been demonstrated that the analytical procedure has a suitable level of precision, accuracy, and linearity. Range is normally expressed in the same units as test results (e.g. percentage, parts per million, etc.) obtained by the analytical method. Range (sometimes referred to as *Linearity of Method*) is evaluated using samples and must encompass the specification range of the component assayed in the drug product.

Raw Data — Raw data are defined as the original record of measurement or observation. Raw data may include, but are not limited to, printed instrument output, electronic signal output, computer output, hand recorded numbers, digital images, hand drawn diagrams, etc. Raw data are proof of the original measurement or observation and by definition cannot be regenerated once collected.

Raw Material — Any ingredient intended for use in the production of intermediates, APIs, or drug products.

Reference Standard — A reference standard is a highly purified compound that is well characterized. It is used as a reference material to confirm the presence and/or amount of the analyte in samples.

Related Compounds — Related compounds are categorized as process impurities, degradants, or contaminants that may be found in finished drug products.

Relative Response Factor (RRF) — The ratio of the response factor of a major component to the response factor of a minor peak. This allows the accurate determination of a minor component without the need for actual standards.

Relative Retention Time (RRT) — The normalization of minor peaks to the parent peak in a chromatogram. The RRT of the parent will be 1.0. Peaks eluting before the parent will have RRTs <1.0. Peaks eluting after the parent will have RRTs >1.0.

Repeatability The variation experienced by a single analyst on a single instrument. Repeatability does not distinguish between variation from the instrument or system alone and from the sample preparation process.

Reporting Limit The level, at or above the LOQ, below which values are not reported (e.g. reported as "<0.05% for a reporting limit = 0.05%"). The reporting limit may be defined by ICH thresholds.

Reprocessing Introducing a previously processed material that does not conform to standards or specifications, back into the process and repeating a step or steps that are part of the established manufacturing process.

Resolution A measure of the efficiency of the separation of two component mixtures. In chromatographic analyses a resolution of >1.5, which means two peaks are separated from each other all the way to the baseline, is desirable.

Response Factor A measuring of the signal generated by a detector normalized to the amount of analyte present. In HPLC it is usually the peak area for a given component (the response) divided by the concentration or mass, which generated that response.

Retain Sample Reserve samples. An appropriately identified sample that is representative of each lot or batch of drug product or API, which is stored under conditions consistent with the product labeling, in the same container-closure system in which the product is marketed or in one that is essentially the same characteristics. Used for testing that may be involved in analyzing complaint samples.

Retrospective Validation Validation of a process for a product already in distribution based upon accumulated production, testing, and control data.

Revalidation The process of partially or completely validating a method or process after changes or modifications have been made to the manufacturing process, analytical methodology, equipment, instrumentation, or other parameter(s), which may affect the quality and composition of the finished product. The USP specifically sites change in synthesis of drug substance, changes in the composition of the drug product, and changes in the analytical procedure.

Robustness The measure of the ability of an analytical method to remain unaffected by small but deliberate variations in method parameters (e.g. pH, mobile phase composition, temperature, instrument settings, etc.) and provides an indication of its reliability during normal usage. Robustness testing is a systematic process of varying a parameter and measuring the effect on the method by monitoring system suitability and/or the analysis of samples. It is part of the formal methods validation process.

Ruggedness A dated term now commonly accepted as intermediate precision. See definition of Intermediate Precision.

Selectivity Selectivity is the ability of the method to separate the analyte from other components that may be present in the sample including impurities. Determination of selectivity normally includes analyzing placebo, blank, media for dissolution, dilution solvent, and mobile phase injections. Also, no chromatographic peaks, such as related compounds, should interfere with the analyte peak, or internal standard peak if applicable. Selective = separate and show every component in the sample.

Signal-to-Noise In chromatography the measure of average baseline noise (e.g. peak-to-peak) to the signal given by an analyte peak. S/N calculations are performed when determining the LOD and LOQ.

SME (Subject Matter Expert) An individual who is consider to be an expert on a particular subject due to a combination of education, training, and experience.

Specification The quality standards (e.g. tests, analytical procedures, and acceptance criteria) provided in an approved application to confirm the quality of the drug substances, drug products, intermediates, raw materials, reagents, and other components including in-process materials.

Specificity Specificity is the ability to assess unequivocally the analyte in the presence of components that may be expected to be present such as impurities, degradation products, and excipients. There must be inarguable data for a method to be specific. Specific = measure only the desired component without interference from other species which might be present; separation is not necessarily required.

Spiked Placebo Preparation of a sample to which known quantities of analyte are added to placebo material. Performed during validation to generate accurate and reproducible samples, which are used to demonstrate recovery from the sample matrix.

Spiking The addition of known amounts of a known compound to a standard, sample, or placebo typically for the purpose of confirming the performance of an analytical procedure or the calibration of an instrument.

Stability Indicating Methodology A validated quantitative analytical procedure or set of procedures that can detect the changes with time in the pertinent properties (e.g. active ingredient, preservative level, or appearance of degradation products) of the drug substance and drug product.

Stability indicating assay. An assay that accurately measures the component of interest (the active ingredient(s) or degradation products) without interference from other degradation products, process impurities, excipients, or other potential interfering substances.

Stability indication profile. A set of procedures or assays that, collectively, detect changes with time although may not do so individually.

Standard and Sample Solution Stability The stability of standards and samples is established under normal bench top conditions, normal storage conditions, and sometimes in the instrument (e.g. an HPLC autosampler) to determine if special storage conditions are necessary, for instance, refrigeration or protection from light. Stability is determined by comparing the response and impurity profile from aged standards or samples to that of a freshly prepared standard and to its own response from earlier time points. Note that these are short-term studies and are not intended to be part of the stability indication assessment or product stability program.

Stressed Studies See Forced Degradation Studies.

System Suitability System suitability is the evaluation of the components of an analytical system to show that the performance of a system meets the standards required by a method. A system suitability evaluation usually contains its own set of parameters; for chromatographic assays, these may include tailing factors, resolution, and precision of standard peak areas, and comparison to a confirmation standard, capacity factors, retention times, theoretical plates, and calibration curve linearity.

Tailing Factor A measure of peak asymmetry. Peaks with a tailing factor of >2 are usually considered to be unacceptable due to difficulties in determine peak start and stop points which complicates integration. Tailing peaks are an indication that the chromatographic conditions for a separation have not been properly optimized.

Test Method An approved, detailed procedure describing how to test a sample for a specified attribute (e.g. assay), including the amount required, instrumentation, reagents, sample preparation steps, data generation steps, and calculations use for evaluation.

Theoretical Plates	A dimensionless quantity used to express the efficiency or performance of a column under specific conditions. A decrease in theoretical plates can be an indication of HPLC column deterioration.
Transcription Accuracy Verification (TAV)	The process where the transcription of data from one location to another is confirmed by a second party. Important with methods validation reports with summary results, which are often transcribed and not linked to raw data.
USP <1225> Category I	One of four method categories for which validation data should be required. Category I methods include analytical methods for quantitation of major components of bulk drug substances or active ingredients (including preservatives) in finished pharmaceutical products. Category I methods are typically referred to as **Assays**.
USP <1225> Category II	One of four method categories for which validation data should be required. Category II methods include analytical methods for determination of **impurities** in bulk drug substances or **degradation compounds** in finished pharmaceutical products. These methods include quantitative assays and limits tests.
USP <1225> Category III	One of four method categories for which validation data should be required. Category III methods include analytical methods for determination of performance characteristics (e.g. **dissolution**, drug release).
USP <1225> Category IV	One of four method categories for which validation data should be required. Category IV methods include analytical methods used as **identification tests**.
Validation	A documented program that provides a high degree of assurance that a specific process, method, or system will consistently produce a result meeting predetermined acceptance criterion.
Validation Characteristics	See Analytical Performance Characteristics.
Validation Parameters	See Analytical Performance Characteristics.

Validation Protocol	A prospective plan that when executed as intended, produces documented evidence that a process, method, or system has been properly validated.
Validation Report	A summary of experiments and results that demonstrate the method is suitable for its intended use, approved by responsible parties.
Verification	The act of reviewing, inspecting, testing, checking, or otherwise establishing and documenting whether items, processes, services, or documents conform to specified requirements.

References

1 21 CFR 211.165. "Testing and release for distribution. (e) The accuracy, sensitivity, specificity, and reproducibility of test methods employed by the firm shall be established and documented. Such validation and documentation may be accomplished in accordance with 211.194(a)(2)."

2 21 CFR 211.194(a)(2). "A statement of each method used in the testing of the sample. The statement shall indicate the location of data that establish that the methods used in the testing of the sample meet proper standards of accuracy and reliability as applied to the product tested. (If the method employed is in the current revision of the United States Pharmacopeia, National Formulary, AOAC International, Book of Methods, or in other recognized standard references, or is detailed in an approved new drug application and the referenced method is not modified, a statement indicating the method and reference will suffice). The suitability of all testing methods used shall be verified under actual conditions of use."

3 EudraLex – Volume 4 – EU Guidelines for Good Manufacturing Practice for Medicinal Products for Human and Veterinary Use, Part 1, Chapter 6: Quality Control.

4 Bliesner, D.M. (2006). *Validating Chromatographic Methods: A Practical Guide*. Wiley-Interscience, John Wiley & Sons.

5 U.S. Department of Health and Human Services Food and Drug Administration Center for Drug Evaluation and Research (CDER) Center for Biologics Evaluation and Research (CBER) (2015). Analytical Procedures and Methods Validation for Drugs and Biologics, Guidance for Industry, July 2015.

6 ICH HARMONISED TRIPARTITE GUIDELINE VALIDATION OF ANALYTICAL PROCEDURES: TEXT AND METHODOLOGY Q2(R1) November 2005.

7 World Health Organization (WHO) (2018). GUIDELINES ON VALIDATION – APPENDIX 4 ANALYTICAL METHOD VALIDATION, September 2018.

8 USP-NF <1224>TRANSFER OF ANALYTICAL PROCEDURES.

9 USP-NF <1225> VALIDATION OF COMPENDIAL PROCEDURES.

10 USP-NF <1226> VERIFICATION OF COMPENDIAL PROCEDURES.

11 USP-NF <620> CHROMATOGRAPHY.

7

Components of the Laboratory Computer Systems Sub Element (LC)

Description of the Laboratory Computer Systems Sub Element

A large variety of analytical instruments, ranging from a simple apparatus to complex computerized systems, is used in the pharmaceutical industry to acquire data that will help ensure that products meet their specifications [1]. However, many, if not most of these instruments are computerized systems; where a computerized system is defined as "A set of software and hardware components which together fulfil certain functionalities" [4]. This chapter, the Laboratory Computer Systems (LC) sub element, addresses laboratory computerized systems by discussing the establishment, operation, and maintenance of a laboratory information technology (IT) infrastructure, functionality, and support system.

Numerous guidance and discussion documents exist, which address computerized systems within the pharmaceutical industry at large [1–5]. However, this chapter is focused on analytical laboratories, and is more limited in scope than a traditional treatise related to IT infrastructure, support, electronic records, etc. If the expertise required to implement and maintain the Laboratory Computer Systems sub element does not exist in-house, laboratory management should seek outside assistance. This may include the site IT and/or Corporate IT functions or a third-party subject matter expert (SME). However, when sourcing laboratory IT support assistance, always ensure that laboratory personnel remain engaged. This is necessary because IT staff often don't possess skills related to laboratory operations, which can lead to challenges in operating the Laboratory Computer Systems sub element in the future.

Laboratory Control System Operations in a GMP Environment, First Edition. David M. Bliesner.

Companion website: www.wiley.com/go/Bliesner/LabControl_GMPEnvironment

Contents of the Sub Element

The Laboratory Computer Systems sub element includes at least four individual topics. A laboratory, which is in compliance with Current Good Manufacturing Practices (CGMPs), should address each of these topics (as applicable). These topics include (i) laboratory IT system administration and procedures, (ii) laboratory computer hardware, (iii) laboratory software, and (iv) laboratory IT-related services. These four topics are described in the succeeding text along with some suggestions on what items should be addressed in each topic.

It should be noted that although similar, complimentary, and sometimes overlapping, the LC sub element differs from the Data Governance and Data Integrity (DI) sub element. Specifically, the LC sub element focuses on Laboratory IT infrastructure and support. The DI sub element focuses on the procedural, personnel, and process aspects, which insure complete, consistent, and accurate data and records throughout the data lifecycle.

Table 7.1 Components of the laboratory computers system sub element.

#	Laboratory computer systems sub element topic	Notes and clarifications
7.1	***Laboratory IT System Administration and Procedures***	
7.1.1	**Roles and Responsibilities**	Laboratory Information Technology (IT) requires a unique set of skills, which includes not only an understanding of IT in general, but also a working knowledge of laboratory operations and the associated computerized systems which function within the laboratory environment.
7.1.1.1	Do the following positions exist to help support administration of the Laboratory Computer Systems sub element:	
7.1.1.1.1	Lab IT Specialist?	This position is often filled by analyst or former analysts who have a strong interest and background in IT.
7.1.1.1.2	Analysts	Senior analysts, who are actively involved in testing at the bench, need to be consulted in any matters related to the Laboratory Computers sub element.

Table 7.1 (Continued)

#	Laboratory computer systems sub element topic	Notes and clarifications
7.1.1.1.3	Site IT	At the bare minimum Site IT should be involved in a consulting and oversight role.
7.1.1.1.4	Corporate IT	
7.1.1.1.5	Third-Party/Outsourced IT/Consultants	Equipment vendors can sometimes be good resources for highly computerized systems like HPLCs, Mass Spectrometers, etc.
7.1.1.2	Do job descriptions and training records exist for the positions described earlier?	As with anyone working in a GMP environment, each individual providing laboratory IT support must possess the qualification, training, and background to do so.
7.1.2	**Computerized System Inventory**	
7.1.2.1	Does an accurate and up-to-date inventory of all computerized systems exist?	This may include both the hardware and corresponding software as appropriate.
7.1.2.2	Does the inventory of all computerized systems include the following?	
7.1.2.2.1	Description of ownership.	
7.1.2.2.2	Description of who supplied or developed the system. Supplier/developer.	
7.1.2.2.3	Description of system functionality.	
7.1.2.2.4	Validation status.	
7.1.3	**IT Service Agreements**	Quality Agreements (Service Level Agreements) should exist for all internal, external, and third-party organizations that provide laboratory IT support. Key roles, responsibilities, and procedures need to be clearly documented in these agreements.
7.1.3.1	Do laboratory IT service agreements consider the following?	
7.1.3.1.1	Hardware and software installation.	
7.1.3.1.2	Hardware and software configuration.	
7.1.3.1.3	Hardware qualification.	

(Continued)

Table 7.1 (Continued)

#	Laboratory computer systems sub element topic	Notes and clarifications
7.1.3.1.4	Software validation.	
7.1.3.1.5	Off-site retention of legacy hardware and software.	
7.1.3.1.6	Network security monitoring.	May be performed via remote access.
7.1.3.1.7	Housing and maintenance of remote servers?	
7.1.3.1.8	Disaster recovery planning and drills.	
7.1.3.1.9	Other services as defined and agreed to in writing.	
7.1.4	**Procurement of Laboratory Computer Hardware, Software, and Systems**	
7.1.4.1	Does a procedure exist, which addresses procurement of laboratory IT hardware, software and computerized systems and does it include the following points:	
7.1.4.1.1	Budgeting and budgetary considerations for laboratory computer systems and IT infrastructure?	
7.1.4.1.2	A needs assessment?	
7.1.4.1.3	Vendor qualification and selection process?	
7.1.4.1.4	Regulatory requirements including:	
7.1.4.1.4.1	Compliance with FDA Part 11 requirements?	Part 11, Electronic Records; Electronic Signatures.
7.1.4.1.4.2	Compliance with EU Annex 11 requirements?	Volume 4 Good Manufacturing Practice Medicinal Products for Human and Veterinary Use Annex 11: Computerized Systems.
7.1.4.1.4.3	Hardware and software installation? Hardware and software configuration?	
7.1.4.1.4.4	Hardware qualification?	
7.1.4.1.4.5	Software validation?	

Table 7.1 (Continued)

#	Laboratory computer systems sub element topic	Notes and clarifications
7.1.5	**Validation of Laboratory Computerized System Applications**	Application = Software.
7.1.5.1	Does a procedure exist, which addresses validation of computerized system applications?	
7.1.5.2	Does the procedure for validating computerized system applications include the following sections:	
7.1.5.2.1	Developing User Requirements Specifications?	Commonly just referred to as a URS.
7.1.5.2.2	Creation of a QA reviewed and approved protocol?	
7.1.5.2.3	Creation of procedures for using the application after the validation?	
7.1.5.2.4	Review and retention of all output and records created during validation of the application?	
7.1.5.3	Does the procedure for validating computerized system applications require confirmation of the following:	
7.1.5.3.1	Data capture?	
7.1.5.3.2	Data analysis?	
7.1.5.3.3	Reporting of results?	
7.1.5.3.4	System security?	
7.1.5.3.5	Access controls?	
7.1.5.3.6	Audit trail functionality?	
7.1.5.3.7	Documenting and reporting protocol deviations?	
7.1.5.3.8	Documenting and reporting failure to meet URSs and/or protocol acceptance criteria?	
7.1.5.3.9	Reporting successful validation of the software application as appropriate?	
7.1.5.3.10	Retaining change control records?	

(Continued)

Table 7.1 (Continued)

#	Laboratory computer systems sub element topic	Notes and clarifications
7.1.6	**Qualification of Laboratory IT Infrastructure**	These efforts should be spearheaded by the Site or Corporate IT Departments with active participation by laboratory IT staff and/or laboratory management.
7.1.6.1	Does a procedure exist, which addresses qualification of laboratory IT infrastructure?	
7.1.6.2	Does the procedure for qualifying laboratory IT infrastructure address the following points?	
7.1.6.2.1	Confirmation that the utilities/facility/environment are satisfactory and meet supplier specified needs for the installed computerized systems?	
7.1.6.2.2	Ensure computerized systems are installed per vendor requirements?	
7.1.6.2.3	Verification that all systems communicate with each other as appropriate?	
7.1.6.2.4	Verify that all data storage and retrieval activities function as appropriate?	
7.1.6.2.5	Verify Audit Trail functionality as appropriate?	
7.1.6.3	Is the qualification of the laboratory IT infrastructure documented and reported appropriately, including:	
7.1.6.3.1	Proof that the QA approved protocol(s) were followed as written?	
7.1.6.3.2	Documenting and reporting failure to meet any protocol acceptance criteria?	
7.1.6.3.3	Reporting successful qualification of the laboratory IT infrastructure?	
7.1.6.3.4	Retention of all records associated with the IT infrastructure qualification?	
7.1.6.3.5	Retaining change control records?	

Table 7.1 (Continued)

#	Laboratory computer systems sub element topic	Notes and clarifications
7.1.7	**Computerized System Accounts**	
7.1.7.1	Does a procedure exist, which addresses the following with respect to computerized system accounts?	
7.1.7.1.2	New account creation?	This includes administrative network accounts, accounts for networked instrumentation, and stand-alone instrument accounts.
7.1.7.1.3	Account modification?	Includes expanding or restricting access as appropriate, which may be a result of promotions, demotions, new job assignments, etc.
7.1.7.1.4	Account closure?	This includes personnel resignation, termination, reassignments, etc.
7.1.8	**Data Structure**	Data structure is defined as a systematic naming convention for electronic data files, data folders, and their associated logical network (or local workstation drive) paths.
7.1.8.1	Does a procedure exist, which addresses data structure?	The goal of creating a data structure is to facilitate the ability to logically track reported results back to raw data in a timely and accurate fashion.
7.1.8.2	Does the procedure on data structure address the following?	
7.1.8.2.1	A standardized data file naming system for files residing on a network?	This often includes some link to the instrument on which the data were acquired, the date it was acquired, file version number, etc. File names may vary from instrument to instrument and where the data reside.
7.1.8.2.2	A standardized data folder naming system for folders residing on a network?	This includes a description of the path to reach the data folder and files, which reside within the folders.

(Continued)

Table 7.1 (Continued)

#	Laboratory computer systems sub element topic	Notes and clarifications
7.1.8.2.3	A standardized data folder and file naming system for non-networked, stand-alone instrument workstations?	All efforts should be made to eliminate stand-alone workstations attached to instruments. From a compliance standpoint, regulators regard data acquired, and stored on stand-alone workstations as vulnerable to data loss and manipulation.
7.1.8.3	Is the ability to create and name files and folders on network drives and stand-alone workstations restricted?	See CRUD matrix discussion in the Computerized System Security, in the succeeding text.
7.1.9	**Computerized System Security**	
7.1.9.1	Does a procedure exist, which discusses laboratory computer systems access control?	
7.1.9.2	Does the procedure related to laboratory computer systems access control address or discuss the following points?	
7.1.9.2.1	CRUD matrix creation and management?	This matrix assigns the following privileges to the laboratory computer systems users: C – Creation R – Read U – Update D – Delete
7.1.9.2.2	Password management?	This includes (i) Issuance, (ii) Scheduled changes (e.g. updates), (iii) Restrictions on re-use of a previously used or existing password, (iv) Reset policy and procedures, (v) Lockout policy, (vi) Restrictions on password sharing, and (vii) Discussion of consequences of violating password procedures.
7.1.9.2.3	Workstation time-outs?	Workstations should be set to automatically lock access after a reasonably brief time.
7.1.9.2.4	Restricted access to DVD drives?	
7.1.9.2.5	Restricted access to flash drive ports?	

Table 7.1 (Continued)

#	Laboratory computer systems sub element topic	Notes and clarifications
7.1.9.2.6	Restrictions on access to the Internet?	
7.1.9.2.7	Policy on locking access to workstations when leaving them unattended?	
7.1.9.2.8	Operating System clock and calendar modification restrictions?	
7.1.9.2.9	Use of electronic signatures?	This may be a separate SOP all unto itself.
7.1.9.3	Does a procedure exist, which addresses data storage, security, and retention?	
7.1.9.4	Does the procedure, which addresses data storage, security, and retention discuss or address the following data and record retention topics?	
7.1.9.4.1	Types of data and records which must be retained?	
7.1.9.4.2	Length of retention time?	Industry-wide, 10 yr are considered the shortest retention period, although many companies are now making data and record retention permanent.
7.1.9.4.3	Retaining all software to allow reconstruction of all GMP activities related to the data and records?	
7.1.9.4.4	Retaining all hardware to interface with the software to allow reconstruction of all GMP activities related to the data and records?	
7.1.9.5	Does the procedure, which addresses data storage, security, and retention discuss or address the following physical security topics:	
7.1.9.5.1	Restricting access to laboratory spaces?	
7.1.9.5.2	Restricting access to server rooms?	Access should be restricted to laboratory IT staff and a limited number of managers.
7.1.9.5.3	Off-Site storage of paper and non-electronic (e.g. physical) media?	Non-electronic records may include DVDs, tapes, flash drives, etc.

(Continued)

Table 7.1 (Continued)

#	Laboratory computer systems sub element topic	Notes and clarifications
7.1.9.5.4	Off-Site storage of electronic media?	Electronic data and records stored on a remote or off-site server.
7.1.9.6	Does the procedure, which addresses data storage, security, and retention discuss or address the following electronic security topic?	
7.1.9.6.1	Use of uninterruptable power sources for critical systems?	Uninterruptable power sources are often used at the bench and associated with instruments such as HPLCs. The purpose of these devices is to prevent unplanned instrument shutdown during testing, when short-term power disruptions occur. They also give the analysts time to halt analyses in a controlled fashion in the event that the power outage becomes longer term.
7.1.9.6.2	Use of backup generators for longer-term power failures?	
7.1.9.6.3	Use of surge protectors for all instruments attached to electrical power?	This will help prevent damage in "power pops" or when surges occur.
7.1.9.6.4	Installation of anti-virus software on server systems and stand-alone instruments?	
7.1.9.6.5	Use of firewalls to protect networked systems?	A firewall is a network security system that monitors and controls incoming and outgoing network traffic based on predetermined security rules [10]. Firewalls typically establish a barrier between a trusted internal network and untrusted external network, such as the Internet [10].
7.1.9.6.6	Enforcement and monitoring of remote instrument access?	Instruments such as HPLCs can be connected to local networks and/or the Internet, giving analysts the ability to monitor chromatographic runs and control the instrumentation remotely. These configurations are not without risks and there needs to be a written policy concerning their use.

Table 7.1 (Continued)

#	Laboratory computer systems sub element topic	Notes and clarifications
7.1.9.6.7	Data breach incident management?	
7.1.9.7	Does a procedure exist, which addresses disaster prevention, recovery, and plans for business continuity?	
7.1.9.7.1	Does the procedure, which addresses disaster prevention, recovery, and plans for business continuity, have provisions for:	
7.1.9.7.2	Data and record backups?	Back-ups of all relevant data should be done. Backups can be done incrementally, periodically, or continuously. Backups are designed as a short-term insurance policy to facilitate disaster recovery, while an archive is designed to provide ongoing access to historical information. The integrity and accuracy of backup data and the ability to restore the data should be checked during validation and monitored periodically [1].
7.1.9.7.3	Data and record archival?	Archival is long-term data and record storage.
7.1.9.7.4	Restoration of data and records?	
7.1.9.7.5	Comprehensive plans for transference of Site activities to an alternate location within the company or to an outsourcing service provider?	
7.2	***Laboratory Computer Hardware***	
7.2.1	Does an accurate and up-to-date inventory of instrument interfaced computerized system hardware exist?	Note: This can also be part of the inventory listed in 7.1.2 "Computerized System Inventory."
7.2.1.1	Does the system hardware inventory include the following?	
7.2.1.1.1	Associated equipment/instrument identification/inventory number?	

(Continued)

Table 7.1 (Continued)

#	Laboratory computer systems sub element topic	Notes and clarifications
7.2.1.1.2	Associated equipment/instrument description?	
7.2.1.1.3	A description of interfaced computer hardware such as a workstation, laptop, tablet, etc.?	
7.2.1.1.4	The purpose of the computer hardware associated with the equipment/instrument such as instrument control, data collection and retention, or data processing and reporting?	
7.2.1.1.5	The qualification status of the instrument and computer hardware?	
7.2.2	Does a description of the physical arrangements, data flows, and other system interactions for network hardware exist?	Network hardware may include servers, routers, cabling, hubs, gateways, and wireless access points.
7.2.3	Does a laboratory administrative computerized inventory exist?	This inventory may include a listing of (i) Laboratory desktop or office computers and their locations, (ii) Printers and their locations, (iii) Copy machines and their locations, and (iv) All-In-One devices (e.g. Copier, scanner, fax, printer) and their locations.
7.3	***Laboratory Software***	
7.3.1	Does an accurate up-to-date inventory of all software installed on computerized systems exist?	Examples of types of laboratory software includes (i) Instrument control, (ii) Data acquisition, (iii) Data processing and reporting, (iv) Laboratory Information Management Systems (LIMSs), (v) Firmware, (vi) Document Management, (vii) Training Management, (viii) Equipment Maintenance and Calibration, and (ix) Office productivity software such as spreadsheets, etc. All types of software used for GMP operations must be validated as appropriate, including document management, training management and maintenance, and calibration databases.

Table 7.1 (Continued)

#	Laboratory computer systems sub element topic	Notes and clarifications
7.3.2	Does the software inventory include the following?	
7.3.2.1	A description of the hardware on which the software is installed?	
7.3.2.2	The name, title, and department of the assigned software owner?	
7.3.2.3	Vendor who supplied the software?	
7.3.2.4	A description of the functionality of the software?	
7.3.2.5	The software installation date?	
7.3.2.6	The software validation status?	
7.3.2.7	Status of any needed updates, patches, or revisions?	
7.3.2.8	Date of next planned update or upgrades?	
7.3.3	Is there a means to ensure that personnel who use the software are properly trained on its use?	
7.3.4	Is someone assigned the duties of ensuring that the installed software is properly maintained?	This includes purchasing and installing upgrades, installing patches, and upgrading firmware if appropriate. Installed software may have an installed-on-hardware shelf-life, and may require a periodic clean re-installation.
7.3.5	Is someone assigned the duties of reinstalling software following hardware failure, hardware fixes, or hardware upgrades?	
7.3.6	Is a LIMSs used for data management?	
7.3.6.1	If a LIMS system is used for data management have the following points and/or topics been taken into consideration:	
7.3.6.1.1	Has the LIMS system been validated?	
7.3.6.1.2	Has the LIMS system validation been fully documented and the documentation is available for review?	

(Continued)

Table 7.1 (Continued)

#	Laboratory computer systems sub element topic	Notes and clarifications
7.3.6.1.3	Is the LIMS system operated and maintained by a qualified laboratory IT staff member or an LIMS IT support organization?	
7.3.6.1.4	Is the LIMS system maintained and updated under a formal change control procedure?	
7.3.6.1.5	Have laboratory personnel been trained on LIMS data entry procedures and are the training records available for review?	
7.3.6.1.6	Are there automated data entry and extraction points in the LIMS system, and have the following links been validated (as appropriate)?	
7.3.6.1.7	Bar code readers?	
7.3.6.1.8	GC workstations?	
7.3.6.1.9	HPLC or UPLC workstations?	
7.3.6.1.10	Balances?	
7.3.6.1.11	Titrators?	
7.3.6.1.12	Other components?	
7.3.6.2	Are LIMS data entries made or enabled only by authorized personnel using username and password protections?	
7.3.6.3	Are 21 CFR Part 11 compliant electronic signatures used for when reviewing and approving LIMS entries?	
7.3.7	Are spreadsheets used in the laboratory to perform CGMP operations?	
7.3.7.1	If spreadsheets are used in CGMP operations in the laboratory, have the following topics with respect to their use been addressed?	Spreadsheets are ubiquitous in many laboratories. However, they need to be used in a controlled fashion.
7.3.7.1.1	Are the spreadsheets validated for their intended use?	

Table 7.1 (Continued)

#	Laboratory computer systems sub element topic	Notes and clarifications
7.3.7.1.2	Are the validated sheets change-protected and unalterable during use?	
7.3.7.1.3	Are spreadsheets modified under the change control system?	
7.3.7.1.4	Are there procedures in place which require the archival of versioned spreadsheets?	
7.4	***Laboratory IT Related Services***	
7.4.1	Is the process of sourcing and selecting laboratory services codified in a procedure?	Types of services may include (i) Subscriptions to scientific and compliance databases (e.g. USP-NF online, ScienceDirect, etc.), (ii) Remote backup and archival services, (iii) Network monitoring services, GMP Training services, etc. As with any outsourcing activity, a Quality Agreement between the vendor and the organization is required [8].
7.4.2	Is someone assigned the duties of managing laboratory IT related services?	This may include soliciting feedback from laboratory personnel as to their needs for services, scheduling updates, periodic renewals, and unplanned purchases and updates/upgrades. The assigned duties should also include a budgeting component.

Tools and Templates

The following is provided in electronic format as an Example Tool and Template, which can be found in the Chapter 7 Appendix (www.wiley.com/go/Bliesner/LabControl_GMPEnvironment):

- Topics in Table 7.1 "Components of the laboratory computer systems sub element" in checklist format for use in evaluating, upgrading, or establishing your own laboratory computer systems sub element.

Glossary

21 CFR Part 11	Title 21, Code of Federal Regulations, Part 11. Part 11 established the criteria under, which electronic records and signatures will be considered equivalent to paper records and hand-written signatures in manufacturing processes regulated by the FDA.
Application	Software installed on a defined platform/hardware providing specific functionality.
Archive	Long-term, permanent retention of completed data and relevant metadata in its final form for the purposes of reconstruction of the Good Manufacturing Practice (GMP) process or activity.
Audit Trail	Audit trails are metadata that are a record of GMP critical information (for example, the change or deletion of GMP relevant data), which permit the reconstruction of GMP activities.
Back-Up	A copy of current (editable) data, metadata, and system configuration settings (e.g. variable settings which relate to an analytical run such as instrument operating conditions) maintained for the purpose of computerized system failure.
Change Control	A formal system by which qualified representatives of appropriate disciplines and backgrounds review proposed or actual changes that might affect the validated status of facilities, systems, equipment, or processes. The intent is to determine the need for action to ensure and document that the system is maintained in a validated state.
Commercial off-the-Shelf Software (COTS)	Software commercially available, whose fitness for use is demonstrated by a broad spectrum of users.

Computer Hardware	Various pieces of equipment in the computer systems, including the central processing unit, the printer, ether net card, screen, and other related apparatus.
Computer System (Computerized System)	Computer hardware components assembled to perform in conjunction with a set of software programs, which are collectively designed to perform a specific function or group of functions. Desktop, laptops, and workstations are examples of computer systems.
Data	All original records and true copies of original records, including source data and metadata and all subsequent transformations and reports of these data, which are generated or recorded at the time of the GMP activity or process and allow full and complete reconstruction and evaluation of the GMP activity. Data may be contained in paper records (such as worksheets, forms, and logbooks), electronic records and audit trails, photographs, audio- or video-files or any other media whereby information related to GMP activities is recorded.
Data Integrity	The extent to which all data are complete, consistent, and accurate throughout the data lifecycle.
Electronic Signature	The computer data compilation of any symbol or series of symbols executed, adopted, or authorized by an individual to be the legally binding equivalent of the individual's hand-written signature.
Firmware	A software program permanently recorded in a hardware device.
IT Infrastructure	The hardware and software such as networking software and operation systems, which makes it possible for the application to function.

Life Cycle	All phases in the life of the system from initial requirements until retirement including design, specification, programming, testing, installation, operation, and maintenance.
Network	(i) An interconnected or interrelated group of nodes. (ii) An interconnected communications facility. A Local Area Network (LAN) is a high bandwidth (allowing a high data transfer rate) computer network operating over a small area such as an office or group of offices.
Operating System	A set of software programs provided with a computer that function as the interface between the hardware and the applications program.
Process Owner	The person responsible for the business process.
Raw Data	Any worksheets, records, memoranda, notes, or exact copies thereof that are the result of original observations and activities and that are necessary for the reconstruction and evaluation of a work project, process, or study report, etc. Raw data may be hard/paper copy or electronic but must be known and defined in system procedures.
Security	The protection of computer hardware and software from accidental or malicious access, use, modification, destruction, or disclosure. Security also pertains to personnel, data, communications, and the physical protection of computer installations.
Standalone System	A self-contained computer system, which provides data processing, monitoring, or control functions but which is not embedded within automated equipment. This is contrasted with an embedded system, the sole purpose of which is to control a particular piece of automated equipment.
System Owner	The person responsible for the availability and maintenance of a computerized system and for the security of the data residing on that system.

Third Party	Parties not directly managed by the holder of the manufacturing and/or import authorization.
User Requirements Specification (URS)	The set of owner, user, and engineering requirements necessary and sufficient to create a feasible design meeting the intended purpose of the system.

References

1. EudraLex Volume 4, Good Manufacturing Practice Medicinal Products for Human and Veterinary Use, Annex 11: Computerised Systems, 30 June 2011.
2. Pharmaceutical Inspection Convention Pharmaceutical Inspection Co-operation Scheme, PI 041-1 (Draft 2) Good Practices for Data Management and Integrity in Regulated GMP/GDP ENVIRONMENTS, 10 August 2016.
3. Pharmaceutical Inspection Convention Pharmaceutical Inspection Co-operation Scheme, Good Practices For Computerised Systems In Regulated "GXP" Environments, PI 011-3 25, September 2007.
4. Pharmaceutical Inspection Convention Pharmaceutical Inspection Co-operation Scheme, PE 009-14 Annexes, Annex 11, Computerized Systems. 01 July 2018.
5. USP <1058> ANALYTICAL INSTRUMENT QUALIFICATION.
6. EudraLex Volume 4, Good Manufacturing Practice Medicinal Products for Human and Veterinary Use, Annex 15: Qualification and Validation, 30 March 2015.
7. USFDA (2003). Guidance for Industry. Part 11, Electronic Records; Electronic Signatures-Scope, and Application, August 2003.
8. USFDA (2016). Guidance for Industry. Contract Manufacturing Arrangements for Drugs: Quality Agreements, November 2016.
9. Contract Pharma Magazine (2019). Glossary of Pharma and Biopharma Industry Terms, April, 2019.
10. Boudriga, N. (2010). *Security of Mobile Communications*, 32–33. Boca Raton, FL: CRC Press. ISBN: 978-0849379420.

8

Components of the Laboratory Investigations Sub Element (LI)

Background and Regulatory History of Out-of-Specification Investigations

Laboratory testing is required by the CGMP regulations in Sections 211.160 and 211.165, in order to confirm that components, containers and closures, in-process materials, and finished products conform to specifications, including stability specifications. Testing associated with analytical and process validation efforts as well as active pharmaceutical ingredients (APIs) is also covered by the CGMPs [1, 5].

The regulations also require that an investigation be conducted whenever an Out-of-Specification test result (OOS) is obtained (§ 211.192) [6]. For purposes of this chapter, the term OOS result includes all test results that fall outside the specifications or acceptance criteria established in drug applications, drug master files (DMFs), official compendia, or by the manufacturer. The term also applies to all in-process laboratory tests that are outside of established specifications.

Over time, the manner and requirements for conducting investigations have evolved. Figure 8.1 presents a historical perspective and description of this evolution.

Description of the Laboratory Investigations Sub Element

The purpose of laboratory investigations is to determine the root cause of an OOS result. The source of the OOS result should be identified either as an irregularity of the laboratory testing or an aberration of the manufacturing process. Even if a batch is rejected based on an OOS result, the investigation is necessary to determine if the result is associated with other batches of the same drug product or other products. Batch rejection does not negate the need to perform the investigation. The regulations require that a written record of the investigation be made, including the conclusions and follow-up (§ 211.192). To be meaningful,

Laboratory Control System Operations in a GMP Environment, First Edition. David M. Bliesner.

Companion website: www.wiley.com/go/Bliesner/LabControl_GMPEnvironment

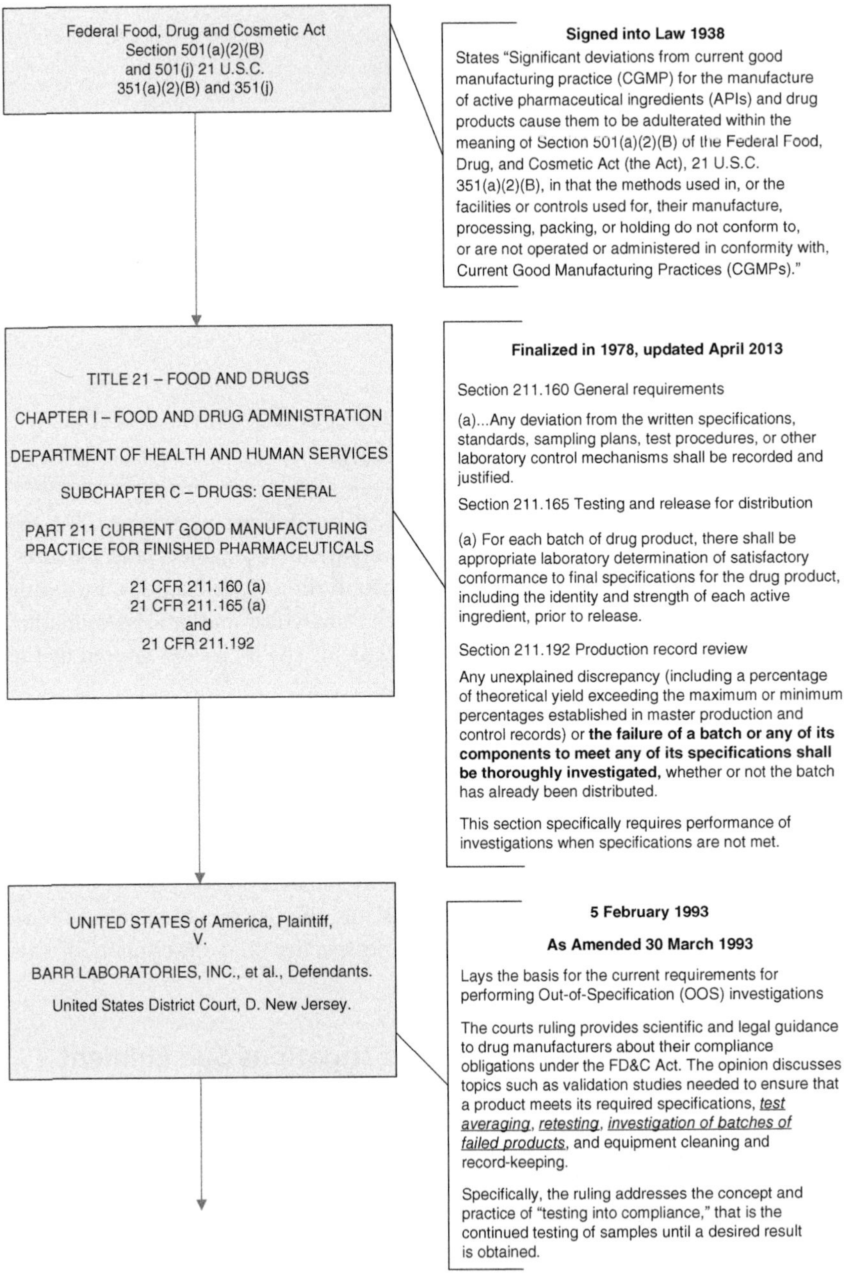

Figure 8.1 The regulatory history of out-of-specification investigations.

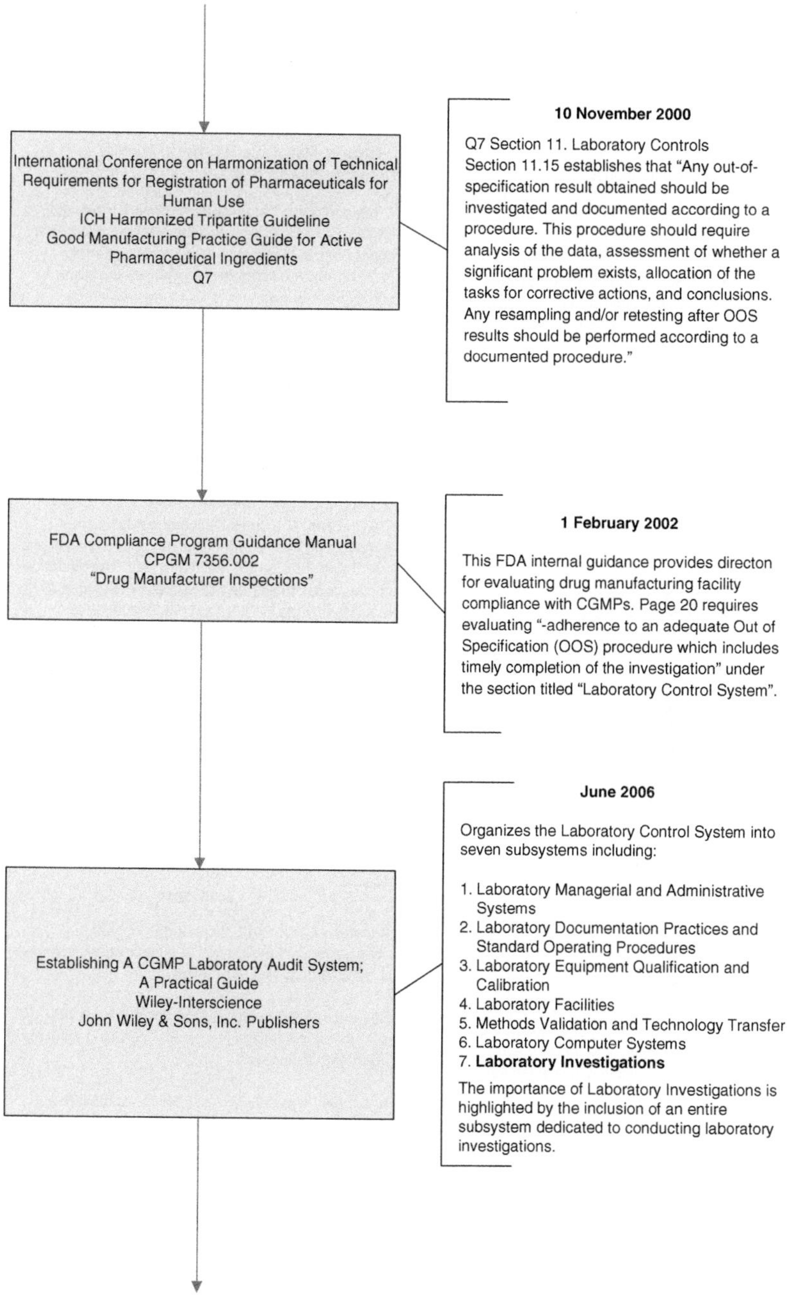

Figure 8.1 (*Continued*)

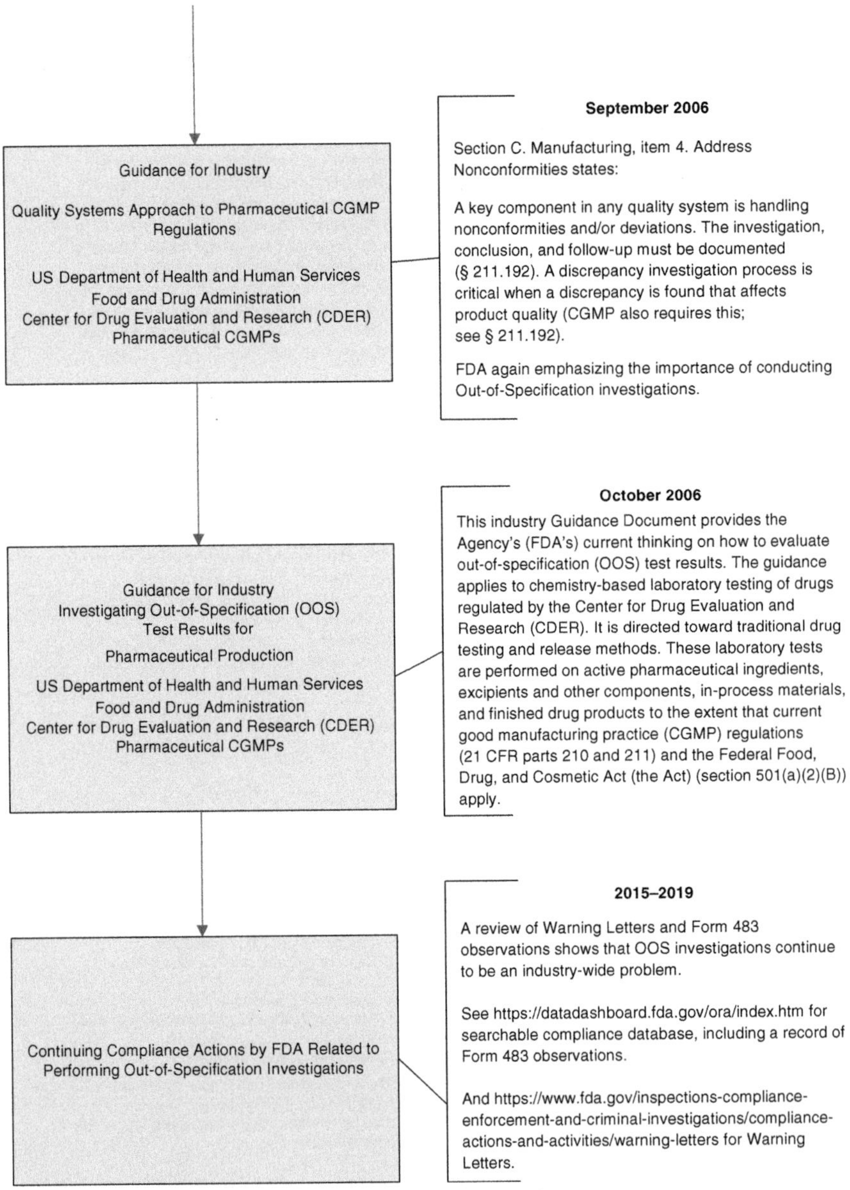

Figure 8.1 (*Continued*)

the investigation should be thorough, timely, unbiased, well-documented, and scientifically sound.

The Laboratory Investigations sub element is designed to systematically identify the root cause of failures, to determine how to correct the problems, and to prevent them from recurring in the future.

Contents of the Sub Element

As indicated in Figure 8.1, despite regulatory requirements and efforts by regulatory agencies such as FDA and MHRA to provide guidance, failure to properly perform laboratory investigations continues to be a challenge for QC testing laboratories [1, 2]. Therefore, the goal of this chapter is to impart a better understanding of the structure and components of laboratory investigations to assist laboratory personnel in improving the quality of their investigations.

In addition to the historical perspective, it is also important to understand the overall structure of investigative process. Figure 8.2 is an adaptation of the United Kingdom's Medicines & Healthcare Products Regulatory Agency's (MHRA) guidance titled "Out of Specification & Out of Trend Investigations" and shows the structure or hierarchy of what components constitute laboratory investigations [2].

It is important to note that laboratory investigations are executed in three phases, namely, Phase 0, Phase I, and Phase II. Table 8.1 describes each of these phases and provides examples of the elements of each phase in addition to typical required actions associated with each phase.

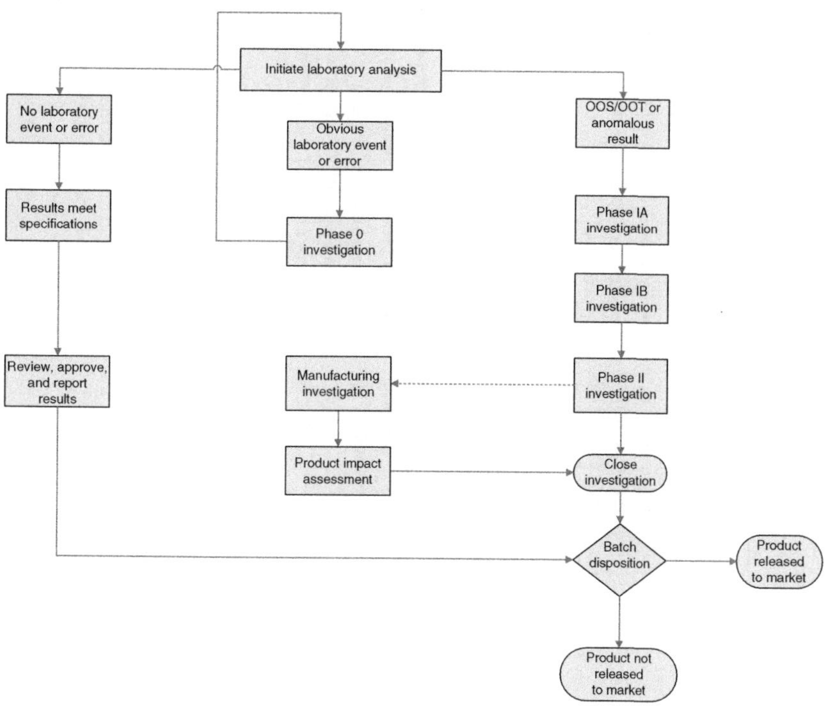

Figure 8.2 The structure of laboratory investigations.

Table 8.1 Description of the phases of a laboratory investigation.

Investigation phase	Description of phase	Elements of the investigation	Required actions	Comments
Phase 0	A Phase 0 investigation is for the identification, documentation, and correction of obvious mistakes, errors, or laboratory events. Phase 0 investigations are **performed in circumstances where no results have yet been generated**.	The following are examples of obvious mistakes, errors, or events associated with Phase 0 investigations: • System suitability failures • Recognition of failure to follow test method • Recognition of an incorrect standard, sample, or reagent weighing • Recognition of a dilution error • Recognition of an incorrect sample, standard or solution preparation • Sample or standard spillage • Recognition of the use of incorrect standards prior to sample injection • Recognition of use of expired solutions prior to sample analysis • Equipment or instrument malfunctions prior to sample analysis • Power failures prior to sample analysis • Aborted or suspended tests due to equipment instability or for long pauses in testing • Errors in file or folder naming conventions	Required actions may include: • Analyst immediately halts all test preparation or testing activities upon recognition of laboratory event, mistake, or error. • Analyst notifies the laboratory Supervisor. • Analyst and Supervisor review all associated equipment, samples, standards documentation, associated with the investigation as appropriate. • Analysts and/or laboratory management document all activities and observations related to the halted test-related or testing activities. • Create a Phase 0 investigation record in the investigation tracking and trending system. • Supervisor reviews and approves the record.	Analyst should not knowingly continue analyses they expect to invalidate at a later time. If samples have been analyzed, a Phase I investigation must be initiated. Performing a periodic review of Phase 0 investigations, especially related to system suitability and equipment failures can prove to be very valuable in mitigating continuing problems with testing. The data from the periodic review may be parsed by method, instrument, analysts, product, etc. and graphed accordingly.

Phase I-A	Phase I investigations are performed in circumstances where OOS, OOT, atypical, aberrant, or anomalous results have been generated. Phase I investigations are divided into two parts: Phase I-A and Phase I-B. A Phase I-A investigation is an initial laboratory investigation **to determine whether there has been a clear or obvious, laboratory error which generated an OOS, OOT, atypical, aberrant, or anomalous result**.	Examples of clear or obvious errors associated with Phase I-A investigations include: • Calculation errors • Power outages • Equipment failures • Testing Errors including, spilling of the sample solution, incomplete transfer of a sample, etc. • Incorrect instrument parameters including setting the detector at the wrong wavelength, incorrect gradient program, etc.	Required actions may include: • *For a calculation error*. Analyst and Supervisor review the error, and both initial and date correction. • *For a power outage*.Analyst and Supervisor document the event, and annotate with words such as "power failure; analysis to be repeated" on all associated and appropriate analytical documentation. • *For an equipment failure*. Analyst and Supervisor document the event, annotate with words such as "equipment failure; analysis to be repeated" and cross reference the equipment maintenance and calibration records. • *For a testing error*. Testing errors may include actions such as spilling of the sample solution, incomplete transfer of a sample, etc. These errors must be documented at the time of their occurrence by the analyst. • *For incorrect instrument parameter*. The Analyst and Supervisor document the event, annotate with words such "incorrect instrument parameter; analysis to be repeated" on all associated and appropriate analytical documentation.	If no error was noted during Phase I-A, then a Phase I-B investigation must take place.

(Continued)

Table 8.1 (Continued)

Investigation phase	Description of phase	Elements of the investigation	Required actions	Comments
Phase I-B	A Phase I-B investigation is performed when an OOS, OOT, atypical, aberrant, or anomalous result has been generated during testing yet **no clear or obvious laboratory error was found during the Phase I-A investigation**. Phase I-B investigations are typically conducted using a laboratory investigation checklist.	Phase I-B investigation checklist items may include the following: • Discussion of the test method with the analyst; confirm analyst training, knowledge and execution of the correct test method including most current version. • Examination of the raw data obtained in the analysis, including chromatograms, spectra, etc. and identify anomalous or suspect information. • Verification that the calculations used to convert raw data values into a final, reportable results and are scientifically sound, appropriate, and correct. • Confirmation that the test method has been validated for its intended use. • Determination that no unauthorized or unvalidated changes have been made to the method. • Confirmation that the performance qualification and calibration status of the instruments is current.	Fully document and preserve records of the laboratory assessment as guided by the checklist items. When clear evidence of laboratory error exists, laboratory testing results should be invalidated. When evidence of laboratory error remains unclear, a Phase II full-scale OOS investigation should be conducted to determine what caused the unexpected results.	Laboratory error should be relatively rare. Frequent errors suggest a problem that might be due to inadequate training of analysts, poorly maintained or improperly calibrated equipment, or careless work. Phase I-B investigations are typically conducted by the analysts or investigator with assistance from the supervisor. Phase I-B investigations are typically restricted to reviewing the analysis itself, the data generated and the instruments and equipment used which generated the OOS, OOT, atypical, aberrant, or anomalous results.

- Determination that appropriate reference standards, solvents, reagents, and other solutions that were used, and are within expiry.
- Examination of retained test solutions.
- If necessary and appropriate direct re-injected solutions to determine if transient equipment malfunction is suspected.
- If appropriate perform further extraction of a dosage unit, to determine if incomplete extraction could invalidate the test results.
- Evaluate the historical performance of the test method to ensure that it is performing according to its intended use.
- Additional checklist items may be added as appropriate.

(Continued)

Table 8.1 (Continued)

Investigation phase	Description of phase	Elements of the investigation	Required actions	Comments
Phase II	A Phase II investigation is a full-scale OOS investigation using a predefined and QA approved investigational plan. A Phase II investigation is initiated when the initial (Phase I) assessment does not determine that laboratory error caused the OOS result and that all test results appear to be accurate.	Phase II investigational plan may include the following: • A requirement to identify and investigate any and all potential problems. • A requirement to review the sampling procedures (e.g. How samples were obtained in manufacturing for QC release testing?). • A requirement to examine the original samples or if necessary, examine resampled material. • A requirement to include all manufacturing sites and contract manufacturers as necessary when conducting the investigation. • A requirement to review production records. • A requirement to perform additional laboratory testing as necessary. • A requirement to extend the investigation to other lots of product as appropriate.	The production review should involve all departments that may be implicated for the OOS including manufacturing, maintenance, process development, and engineering. The documentation and records of the manufacturing process should be fully reviewed to determine the possible cause of the OOS result. If some or all of the manufacturing takes place off-site (e.g. multiple company manufacturing sites or at a contract manufacturing organization), these groups should be included in the investigation. Additional laboratory testing may be required and include: • *Retesting*. When no assignable cause exists for the OOS results. Then a portion of the original sample is retested in an effort to confirm or invalidate the OOS results.	The objectives the Phase II investigation should be to (i) Confirm the OOS result, (ii) Disprove the OOS result, or if confirmed (iii) Identify the root cause of the OOS result and take appropriate corrective and preventative action. Quality Assurance (QA) should lead the Phase II investigation. The investigation should be timely, thorough, and well documented and include the following points: • A clear and concise statement as to why the investigation is being performed.

	• A requirement to retest and replace results as appropriate. • A requirement to evaluate the final results of the investigation, evaluate the quality of the batch and make a decision for batch release or reject the batch. • Guidance with respect to further testing to determine the root or most probable cause if the batch is rejected.	• *Resampling*. When no assignable cause exists for the OOS results but none of the original sample remains. Then it is necessary to obtain additional samples and retest them in an effort to confirm or invalidate the OOS results. The investigation is concluded after the results are interpreted, conclusions are drawn based on the results of the investigation and the application of scientifically sound reasoning. QA will make the decisions with respect to batch disposition.	• A summary of the aspects of the production process which may have caused the problem. • The results of the records review with assignment of the actual or probable cause of the failure. • A review of existing investigation records to determine if the failure has previously occurred. • A description of all corrective and preventive actions.

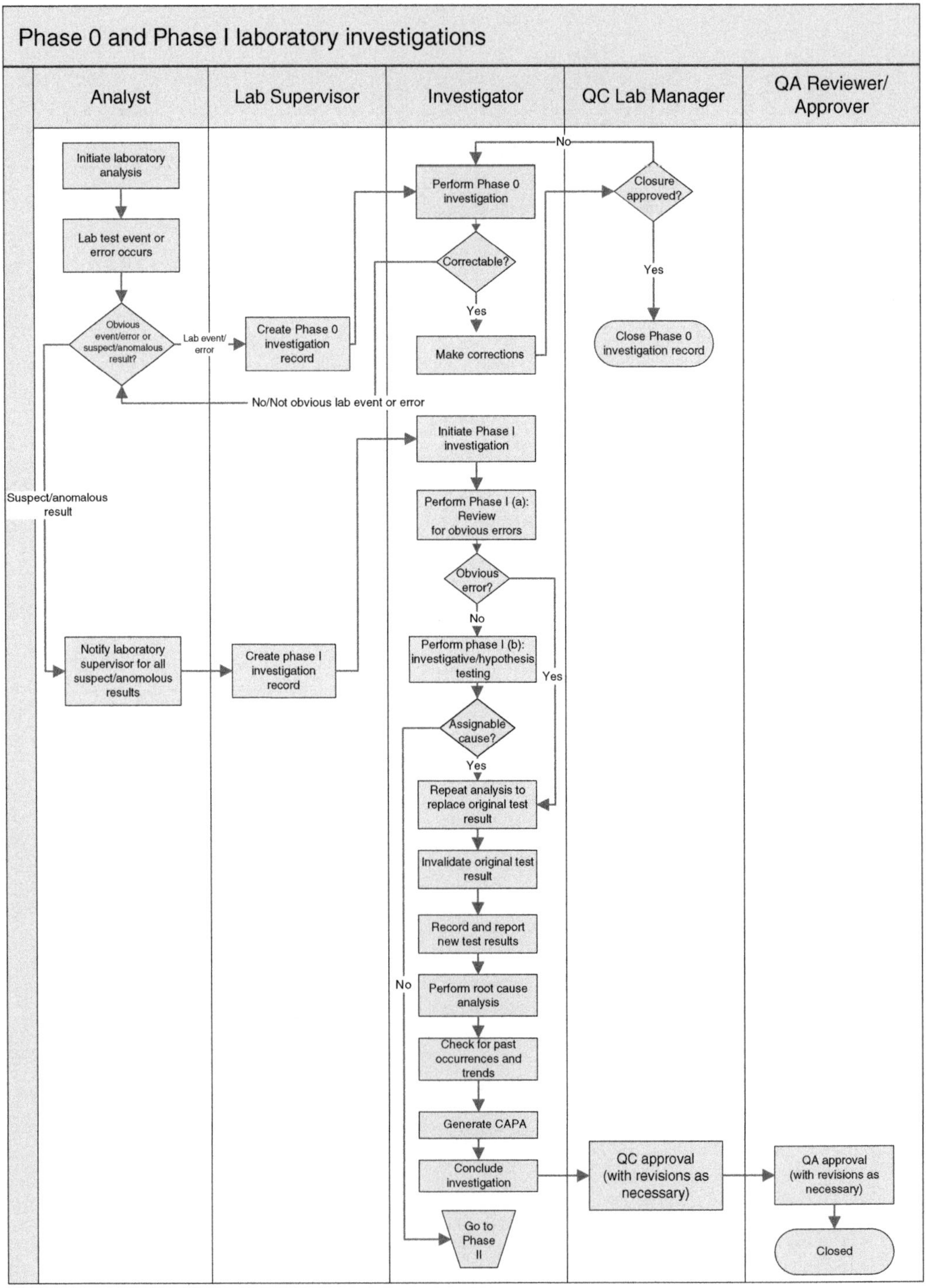

Figure 8.3 A model laboratory investigations work flow diagram.

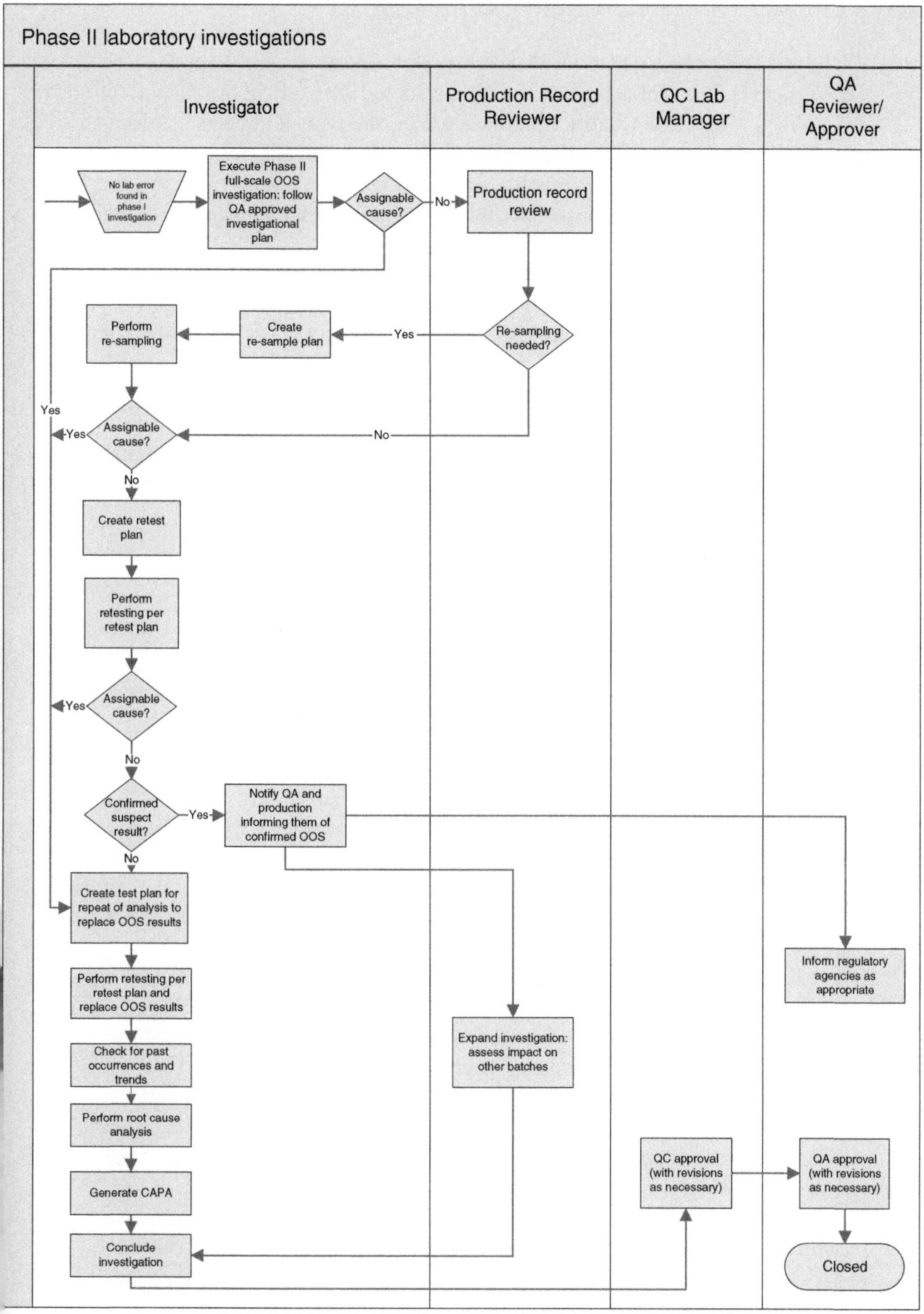

Figure 8.3 (*Continued*)

The exact manner in which each laboratory performs its investigations depends on multiple factors. These include the organizational structure of the laboratory, the defined role of the Quality Assurance group, the type of systems/software, which is used to manage investigations, the types of equipment and instruments used to support the QC testing mission, and other factors. With that said, in general, laboratory investigations are executed in a similar fashion regardless of the laboratory or its testing mission.

Figure 8.3 describes the work flow steps that need to be taken in order perform laboratory investigations in line with the CGMPs and industry guidance. Figure 8.3 graphically describes the steps discussed in Table 8.1 "Description of the phases of a laboratory investigation."

It should be noted that the definitions for terms presented in the work flow diagram are contained in the Glossary section of this chapter.

The procedure for executing laboratory investigations previously shown should be captured and codified in a standard operating procedure.

Common Problems Related to Laboratory OOS Investigations

The following are some common problems related to execution of laboratory OOS investigations:

- Ongoing bias that assumes that most OOS investigations are related to laboratory errors and the final outcome of the investigation will most likely result in release of the product to market.
- Phase II investigations do not prompt a serious and complete review of the production batch record, manufacturing process, and process validation as appropriate.
- A pattern of not determining the root cause/probable cause of the original OOS result with subsequent re-testing and release of product to market.
- Poor analysts' interviews; answers not supported by evidence.
- Investigation is based on speculation not scientific evidence.
- Investigation is not a stand-alone document; can't be read and understood, requires the analysts or investigator describe verbally what happened.
- Belief that test methods have been properly validated or qualified and perform properly for their intended use; where in fact older methods, which have not been subject to lifecycle management, just don't work consistently per their intended use.
- Expendables such as HPLC columns are assumed to be equivalent to the ones on which the method was validated; when in fact expendables like HPLC columns for older analytical methods are no longer available, and have been substituted with non-equivalent products by the vendor, despite what they may say.
- Laboratory investigations are not always performed in a timely manner.

- Confirmed OOS results do not trigger the requirement to determine if other batches are impacted (e.g. expand the investigation).
- Extraneous chromatographic peaks are observed, but not investigated.
- Ambiguous or incorrectly worded analytical test methods; leave execution open to interpretation or leads to faulting execution.
- Ambiguous or incorrectly worded OOS and/or OOT standard operating procedures.
- The OOS SOP is designed around using the OOS database software as opposed to conducting the investigation itself.
- There is no capture, tracking, patterning, and trending of investigations with periodic quality reviews of OOS investigations.
- Phase 0 events such as system suitability failures are not captured, tracked, patterned, or trended.
- Lack of robust corrective and preventive actions (CAPA); no system-based fixes.
- Analysts, supervisors, and managers are not trained and qualified to perform investigations.
- Laboratory equipment not maintained in the state at which the OOS event occurred; next analyst begins using instrument before any data is reviewed, which identifies the OOS.
- Unusual events, suspected or potential problems, are not documented at the time of the event.
- Use of obviously dirty or defective glassware.
- Laboratory testing write-ups are not contemporaneous and steps of the analysis are recorded sometime after the actual steps are executed; for example, analysts perform dilution scheme as directed in the method, but doesn't record the steps until hours (or days!) later or never records them at all. This prevents any effective investigation.
- Investigative forms and checklists are completed without thought or without discussion; literally just checks-in-the-blocks.
- Data not reviewed and approved until well past the time of analysis.
- Additional Investigative "Red Flags"
  - Many root causes assigned as "personnel or training."
  - Many CAPAs are "re-training."
  - Investigations remain open for a long period of time with multiple extensions.
  - Extensions are granted by QA with limited debate or scrutiny.

Tools and Templates

The following is provided in electronic format as Example Tools and Templates, which can be found in the Chapter 8 Appendix (www.wiley.com/go/Bliesner/LabControl_GMPEnvironment):

- A checklist for use in evaluating, upgrading, or establishing your own Laboratory Investigations sub element

- Example-Template laboratory investigations work flow diagram
- Copy of the USFDA Guidance for Industry "Investigating Out-of-Specification (OOS) Test Results for Pharmaceutical Production," October 2006
- Copy of the MHRA Guidance "Out of Specification & Out of Trend Investigations," October 2017
- Copy of the decision titled "UNITED STATES of America, Plaintiff, V. BARR LABORATORIES, INC., et al., Defendants. United States District Court, D. New Jersey." 5 February 1993, As amended 30 March 1993

Glossary

A wide variety of terminology is used in the pharmaceutical industry regarding OOS investigations. The following are some of the more commonly used terms.

It should be noted that care must be taken when adopting any of these terms for use in your organization in order to prevent confusion.

Action Level(s) Levels established for the monitoring of the various systems, which, if exceeded, trigger an investigation into the excursion. Action (or alert) levels are not to be confused with nor used interchangeably with the term specification. Unlike specifications, action level excursions are not tied directly to the quality or profile of a given product but rather demonstrate a deviation from a system's expected performance, the impact of which is evaluated in the investigation.

Alert Level(s) Levels established for the monitoring of the various systems that fall within action levels, yet are outside of the normal performance characteristics of a given system. Alert levels are established to give notice and visibility to potential problem areas.

Atypical/Aberrant/Anomalous/Suspect Result/Unexpected Result A result that initiates a laboratory investigation. The result is still within specification but is inconsistent with previous data, established trends, or other results for the same sample on test. Examples would be chromatograms that show unexpected peaks, unexpected results for stability test point, etc. The result may be out-of-trend or atypical.

Atypical Event An unexpected event, such as power failure or equipment malfunction, that might result in an invalid test result.

Acceptance Criteria	Numerical limits, ranges, or other suitable measures for acceptance of the results of analytical procedures.
Assignable Cause	Clear evidence, scientifically justified and fully documented, which describes the reason for obtaining an OOS or aberrant/anomalous/suspect result.
Laboratory Assignable Cause	A factor in the laboratory, proven through investigation, to be the immediate cause of the OOS, OOT, or atypical results/events, etc.
No Assignable Cause	When no reason for the OOS or aberrant/anomalous/suspect result could be identified.
Deviation	An event, whether planned, unplanned, or unexpected, which constitutes a failure or inability to follow a requirement related to CGMP activities such SOPs. Most likely will require determination of the cause, impact, corrective and preventive actions.
Event	Something that occurs that is outside of what is normal or expected. The Event is only the description of what was observed; it must be considered separate from the Cause of the Event.
Field Alert Report (FAR)	A requirement delineated in 21 CFR 314.81(b)(1)(ii), which requires that information concerning any failure of a distributed batch to meet any of the specifications established in an application, be reported to FDA within three days of discovery.
Hypothesis/ Investigative Testing	Testing performed to help confirm or discount a possible root cause. Examples may include further testing regarding sample filtration, sonication/extraction and potential equipment failures etc. Multiple hypotheses can be explored during investigations. Hypothesis testing may be performed during Phase I and Phase II investigations.
Internal Specification	Action limits within regulatory specifications usually set with more stringent limits than the filed specifications.
Invalidated Test	A test that is considered invalid when the investigation has determined an assignable cause.

Most Probable Cause	Scientifically justified determination of what caused an atypical/aberrant/anomalous/suspect or OOS result.
Out-of-Specification (OOS) Result	A test result that does not comply with the pre-determined acceptance criteria (e.g. filed applications, drug master files, approved marketing submissions, or official compendia or internal acceptance criteria). Test results that fall outside of established acceptance criteria, which have been established in official compendia and/or by company documentation (i.e. Raw Material Specifications, In-Process/Final Product Testing, etc.).
Out-of-Trend (OOT) Result	An Out-of-Trend result is a test result that does not follow the expected trend, either in comparison with previous results generally collected during a stability study. Data from testing raw materials, components, and in-process samples may also yield out-of-trend data. OOT results are not necessarily OOS but do not look like a typical data point.
Out-of-Specification Investigation	The systematic and planned search for the root or most probable cause that generated an out-of-specification result. An OOS investigation should include formal reporting and a description of corrective and preventive actions take.
Outlier	A test result that is statistically different compared with a set of other test results obtained from the same sample or samples from the same lot of material using a validated method.
Outlier Test	A statistical procedure for identifying data that are extreme. Outlier tests may be used to add in OOS investigations; however, Outlier tests should not be used to invalidate data.
Patterning	The use of graphical techniques such as Pareto analysis to identify patterns in data, which have been generated over time. Unlike trending, which looks for changes in a specific measurable value over time, patterning looks for the number of occurrences in data over time.

Phase 0 Investigation	An investigation for the identification, documentation, and correction of obvious mistakes, errors, or laboratory events. Phase 0 investigations are performed where no results have yet been generated. Examples include system suitability failures, recognition of failure to follow the test method, recognition of a dilution error, equipment malfunctions prior to sample analysis, etc.
Phase I Investigation	Phase I investigations are performed in circumstances where OOS, OOT, atypical, aberrant, or anomalous results have been generated. Phase I investigations are divided into two parts. Phase I-A and Phase I-B.
Phase I-A Investigation	A Phase I-A investigation is an initial laboratory investigation to determine whether there has been a clear or obvious laboratory error, which generated an OOS, OOT, atypical, aberrant, or anomalous result.
Phase I-B Investigation	A Phase I-B investigation is performed when an OOS, OOT, atypical, aberrant, or anomalous result has been generated during testing yet no clear or obvious laboratory error was found during the Phase I-A investigation.
Phase II Investigation	A Phase II investigation is a full-scale OOS investing using a QA approved predefined investigational plan and may include hypothesis testing and/or retest plans.
Re-Measurement	Repetition of the measurement of the analytical material that has been completely processed and prepared for direct testing. Re-measurements can consist of re-injection of the final sample preparation or completing the required sample preparation of an intermediate test solution. Re-measurement can occur during Phase I-B and doesn't necessarily require a retest plan.
Re-Examination	Visual assessment of actual test solutions, test units, and glassware used in the original measurements and preparations, which might provide more credibility for laboratory error hypotheses. Usually part of the Phase I-B investigational step.

Reportable Result A final analytical result. This result is appropriately defined in the written approved test method and derived from one full execution of that method, starting from the original sample.

Replicate A single sample preparation analyzed on the same system multiple times (e.g. multiple injections from the same vial on HPLC).

Re-Test/Retesting/ Repeat Testing Additional testing, which is performed when no assignable cause for an OOS was discovered during a Phase I and Phase II investigation. The testing is performed using the material from the original sample composite, if the material has not been compromised and/or is still available. Retesting is performed defined in a QA approved retest plan.

Re-Sample A new sample from the original source (where possible) in the event of insufficient material remaining from original sample composite or proven issue with original sample integrity. Re-sampling should rarely occur.

Root Cause The underlying reason or source of variation causing a problem.

Root Cause Analysis A process to identify the Assignable Cause of an event as well as to identify the established conditions that made the Assignable Cause possible.

Repeat-to-Replace Testing Testing performed on the original sample or re-sampled material to replace invalidated test results.

Reportable Result The final analytical result which meets all acceptance criteria and was obtained by using a validated test method.

Specification A specification is defined as a list of tests, references to analytical procedures, and appropriate acceptance criteria, which are numerical limits, ranges, or other criteria for the tests described. It establishes the set of criteria to which an API (active pharmaceutical ingredients, also known as drug substance), drug product or materials at other stages of its manufacture should conform to be considered acceptable for its intended use.

System Suitability	The checking of an analytical testing system before or during analysis of unknowns to ensure system performance (ICH).
Test Plan	A plan that provides detailed direction for additional testing, which may be needed to support laboratory investigations. Test plans may be created for hypothesis testing, retesting, or repeat-to-replace testing.
Trend	A statistical term referring to the direction or rate of change of a variable(s). (ICH 09).
Trending	The process of evaluating data for atypical observations or excessive variation.

References

1 USFDA (2006). Guidance for Industry. Investigating Out-of-Specification (OOS) Test Results for Pharmaceutical Production, October 2006.

2 Medicines and Healthcare Products Regulatory Agency (2017). (MHRA) Guidance. Out of Specification & Out of Trend Investigations, October 2017.

3 Huxsoll, J.F., Mayer-Sangster, M., Walker, K. et al. (eds.) (2001). *A Laboratory Quality Handbook of Best Practices: Out-of-Specification Results*, Chapter 11. Milwaukee, WI: ASQ Quality Press.

4 UNITED STATES of AMERICA, Plaintiff, V. Barr Laboratories, Inc, et al. Defendants. United States District Court, D., New Jersey. Civil Action No. 92-1744, 5 February 1993, as amended 30 March 1993.

5 ICH HARMONISED TRIPARTITE GUIDELINE GOOD MANUFACTURING PRACTICE GUIDE FOR ACTIVE PHARMACEUTICAL INGREDIENTS, Q7 10 November 2000.

6 USP <1010> ANALYTICAL DATA-INTERPRETATION AND TREATMENT.

7 Bliesner, D.M. (2006). *Establishing A CGMP Laboratory Audit System; A Practical Guide*. Wiley-Interscience, John Wiley & Sons, Inc.

9

Components of the Laboratory Data Governance and Data Integrity Sub Element (DI)

Background

In recent years, the US Food and Drug Administration (FDA) has increasingly observed violations of the current good manufacturing practice (CGMP) regulations with respect to data integrity. This is troubling to FDA in that ensuring the integrity of data is a critically important component of the CGMPs and the pharmaceutical industry's responsibility to ensure the drugs they manufacture are safe and effective and have the quality they purport and of FDA's ability to protect public health. These data integrity-related CGMP violations have led to numerous regulatory actions, including Form 483 observations, Warning Letters, Import Alerts, and in one case, laid the basis for consent decrees (CDs) [1].

FDA defines data integrity as the "…completeness, consistency, and accuracy of data. Complete, consistent, and accurate data should be attributable, legible, contemporaneously recorded, original or a true copy, and accurate (ALCOA)" [1]. However, other regulatory bodies and industry organizations have included a "+" to the acronym where the "+" is Complete, Consistent, Enduring, and Available [2]. See Table 9.1 for a more detail description of the ALCOA+ principals.

Data governance, on the other hand, is best defined as the sum total of arrangements, which provide assurance of data integrity [3]. These arrangements ensure that data, irrespective of the process, format, or technology in which it is generated, recorded, processed, retained, retrieved, and used, will ensure a complete, consistent, and accurate record throughout the data lifecycle. Data governance should be integral to the pharmaceutical quality system. It should address data ownership throughout the lifecycle, and consider the design, operation, and monitoring of processes/systems in order to comply with the principles of data integrity, including control over intentional and unintentional changes to, and deletion of information [3].

As is typical when FDA focuses on a particular aspect of regulatory compliance, the pharmaceutical industry quickly swings into action to try to develop

Laboratory Control System Operations in a GMP Environment, First Edition. David M. Bliesner.

Companion website: www.wiley.com/go/Bliesner/LabControl_GMPEnvironment

Table 9.1 ALCOA+ acronym for data integrity.

Criterion	Meaning
Attributable	Attribute to the person generating the data (MHRA)[a] Who acquired the data originally or performed an action subsequently to it and when?
Legible	Legible (MHRA) Can you read the data together with any metadata or all written entries on paper?
Contemporaneous	Contemporaneous (MHRA) Documented (on paper or electronically) at the time of an activity
Original	Original record or true copy (MHRA) Written observation or printout or a certified copy thereof Electronic record including metadata of an activity
Accurate	Accurate (MHRA) No errors in the original observation(s) No editing without documented amendments/audit trail entries by authorized personnel
Complete	All data from an analysis, including any data generated before a problem is observed, data generated after repeat part or all of the work or reanalysis performed on the sample For hybrid systems, the paper output must be linked to the underlying electronic records used to produce it
Consistent	All elements of the analysis, such as the sequence of events, follow on and data files are date (all processes) and time (when using a hybrid or electronic systems) stamped in the expected order
Enduring	Recorded on authorized media, e.g. laboratory notebooks, numbered worksheets, for which there is accountability or electronic media Not recorded on the backs of envelopes, laboratory coat sleeves, cigarette packets or post-it notes
Available	The complete collection of records can be accessed or retrieved for review and audit or inspection over the lifetime of the record

a) Medicines & Healthcare Product Regulatory Agency.

approaches, which will allay any fears which the agency may harbor. This is in fact is quite true for data governance and data integrity data as well.

For all intents and purposes, data governance and data integrity have become a cottage industry within the regulatory compliance and CGMP consulting community. As a result, a withering number of articles, papers, publications, and presentations have been generated to assist companies to "correct" their data integrity deficiencies. In fact, entire text books and technical reports have been published to address the topic (see Further Reading).

This level of increased scrutiny and required action, for the most part, is a positive development. However, it does come with a price. Namely, one can nearly make a career out of reading, attempting to understand and apply all the principals laid out in these sources of information. Therefore, the purpose of this chapter is to (i) simplify the concepts of data governance and data integrity for easier consumption, (ii) outline a process that organizations can use to comply with the precepts of data governance and data integrity, and (iii) provide some data governance and data integrity tools and templates. This will allow pharmaceutical companies to continue to exercise their missions of manufacturing and releasing pharmaceutical products to market that are safe and effective and have the quality they purport, without getting bogged-down in regulatory compliance minutia. In order for this to happen, some basic precepts concerning data governance and data integrity need to be discussed.

Before proceeding, it should be noted that this chapter addresses data governance and data integrity from a global perspective. That is, data governance and data integrity apply to the entire quality system, not just the lab. However, specifics related to data governance and data integrity within the laboratory control system (LCS) are discussed in detail where appropriate.

Precepts Regarding Data Governance and Data Integrity

1. *Data integrity is not a new concept.*
   - Requirements with respect to data integrity are a major component of the CGMPs, including [1]:
     - Section 211.68 (requiring that "backup data are exact and complete" and "secure from alteration, inadvertent erasures, or loss" and that "output from the computer… be checked for accuracy").
     - Section 212.110(b) (requiring that data be "stored to prevent deterioration or loss").
     - Sections 211.100 and 211.160 (requiring that certain activities be "documented at the time of performance" and that laboratory controls be "scientifically sound").
     - Section 211.180 (requiring that records be retained as "original records," or "true copies," or other "accurate reproductions of the original records").
     - Sections 211.188, 211.194, and 212.60(g) (requiring "complete information," "complete data derived from all tests," "complete record of all data," and "complete records of all tests performed").
     - Sections 211.22, 211.192, and 211.194(a) (requiring that production and control records be "reviewed" and that laboratory records be "reviewed for accuracy, completeness, and compliance with established standards").

   - o Sections 211.182, 211.186(a), 211.188(b)(11), and 211.194(a)(8) (requiring that records be "checked," "verified," or "reviewed").
 - The quality and integrity of data has always been the quality control (QC) laboratories primary mission.
   - o Production's mission is to produce high quality products.
   - o QC's mission is to produce high quality data.
2. *In its simplest description, data integrity is the ability to reconstruct GMP activities.*
   - Every reportable result should be traceable to the original records.
   - Traceability should not be time sensitive: Should be possible even years after the fact.
   - Traceability should not be people dependent: Should be possible by anyone with a basic knowledge of the organization's documentation and data retention procedures.
3. *Data integrity is not just about the LCS or just QC labs.*
   - Most recent data integrity issues have been centered on the QC laboratory.
   - Data integrity issues potentially exist in all six of the FDA quality systems.
   - Manufacturing in particular has a chronic problem with personnel sharing of passwords to access and control manufacturing equipment, which have computer-controlled interfaces.
   - In addition, although formerly somewhat immune to FDA scrutiny, Research and Development laboratories are now becoming "fair game" with respect to regulatory oversight, especially concerning data integrity.
4. *Data governance and data integrity are concepts and processes, not a checklist.*
   - Laboratories struggling with implementing a Data Governance system often think all they need is a checklist.
   - Data governance and data integrity is a sub element of the LCS, and includes all the standard operating procedures (SOPs), process diagrams, trained personnel, etc. that you would expect (and has been previously discussed in this text) for any LCS sub element.
5. *Regulatory and industry guidances for data governance and data integrity can be confusing.*
   - These documents are confusing at best and approach data integrity as a concept not as a system or process.
   - Although the FDA Guidance Documents, MHRA Guidance Documents, and the PIC/S Guidance are useful, they are challenging to apply in real-world scenarios.
   - These documents are a good reference, and a good starting point, but must be applied by personnel who understand the CGMPs and laboratory operations.

6. *The root cause of many data integrity challenges is lack of or misguided leadership.*
   - Leaders (e.g. Corporate management, site management, department management, etc.) ultimately set the tone for compliance with the CGMPs; the wrong message, intentionally or unintentionally, can negatively impact data integrity.
   - Some leadership factors, which can negatively impact data integrity, include:
     - Putting pressure on personnel to release product to market, even if they feel it is not up to quality standards.
     - Displaying a lack of appreciation for the complexity and challenges of LCS operations.
     - Making employees feel that they may be terminated if they ever challenge their leaders.
     - Making employees feel that mistakes will be held against them, and will jeopardize any chance of advancement.
     - Making employees feel they will not be treated with respect if they make honest mistakes.
   - Bottom line: If you're in charge, and the organization has data integrity challenges, go look in the mirror to find the root cause.
7. *Data integrity challenges are not always fraudulent*
   - Challenges to data integrity are most often approached from a standpoint of fraudulent activity; this is unfortunate because it immediately puts anyone involved on the defensive.
   - Lack of integral data is just as likely the result of poor data governance and non-existent systems as it is from fraudulent activity.
8. *Data integrity challenges related to fraud are difficult to detect and to prove.*
   - Trying to uncover fraudulent activity related to data integrity challenges requires specialized skill sets and experience.
   - Investigating fraudulent activity related to data integrity challenges is time consuming and expensive and for all intents and purposes takes on the form of a criminal investigation.
   - The best way to prevent fraudulent activity related to data integrity is to implement systems and encourage the right behaviors.
9. *A well-designed and robust LCS is the best means to ensure the integrity of data.*
   - Implement, operate, and maintain the 10 LCS sub elements.
   - The sub elements, as previously and currently described, include:
     i. Laboratory Managerial and Administrative Systems Sub Element
     ii. Laboratory Documentation Practices and Standard Operating Procedures Sub Element

iii. Laboratory Equipment Sub Element
iv. Laboratory Facilities Sub Element
v. Methods Validation and Method Transfer Sub Element
vi. Laboratory Computer Systems Sub Element
vii. Laboratory Investigations Sub Element
viii. Data Governance and Data Integrity Sub Element
ix. Stability Program Sub Element (as appropriate)
x. General Laboratory Compliance Practices Sub Element

- Laboratories that operate and maintain a compliant LCS, rarely face data integrity challenges.

10. *A close and professional relationship between quality control (QC) and quality assurance (QA) is imperative*
    - It is far better for your QA group to find challenges than to have FDA do so during an inspection or when reviewing your applications.
    - QC needs to encourage and assist QA to develop understanding of laboratory operations including the challenges, which are part of the everyday work world of the laboratory.
    - QA should actively monitor QC operations; observe work at the bench as well as perform document and data review and approval.

Description of the Laboratory Data Governance and Data Integrity Sub Element

The purpose of the data governance system is to implement policies and procedures, which allow, upon request, the means to fully reconstruct good manufacturing practice (GMP) activities by retrieving complete information relating to the production, testing, and release of a manufactured batch of drug or drug product. The remainder of this chapter outlines the possible structure of a data governance system and its key components to ensure the integrity of data. The data governance and data integrity sub element should include at least nine individual components. These include (i) an overarching policy, which establishes and describes the data governance system. (ii) A list of operational procedures, which describe and govern the lifecycle of data (see Figure 9.1). (iii) A list of technical controls, which can be used alone or in combination with procedural controls. (iv) A description of a process to assess and evaluate existing controls over data. (v) Identifying and ranking known risks to data. (vi) Discussions on how data are reviewed and approved for use. (vii) Operating and enhancing the data governance system by conducting

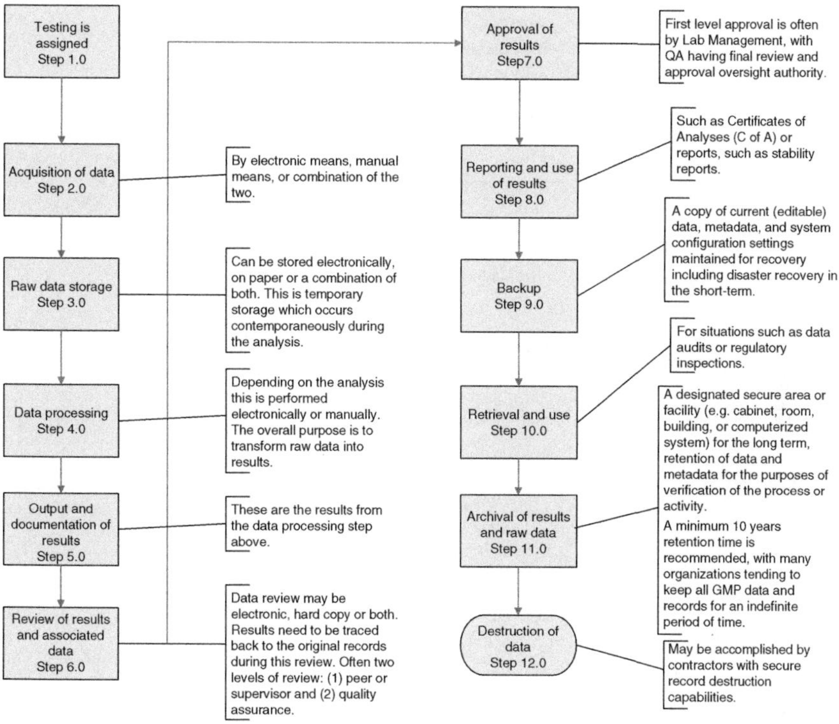

Figure 9.1 The lifecycle of laboratory data.

periodic audits of operations and data. (viii) Enhancing employee awareness and training them on the concepts of data governance and data integrity. (ix) Management oversight of the data governance systems and the integrity of data.

It should be reiterated, that the approach to data governance and data integrity discussed here is not exclusively associated with laboratories, but falls under the scheme of systems for the manufacture of drugs/drug products used by FDA to perform inspections [4]. The scheme of systems includes (i) The Quality System, (ii) The Facilities and Equipment System, (iii) The Materials System, (iv) The Production System, (v) The Packaging and Labeling System, and finally, (vi) The Laboratory Control System.

These systems, and the data governance and data integrity component, exist within a hierarchy of this scheme of systems. This hierarchy is presented graphically in Figure 9.2.

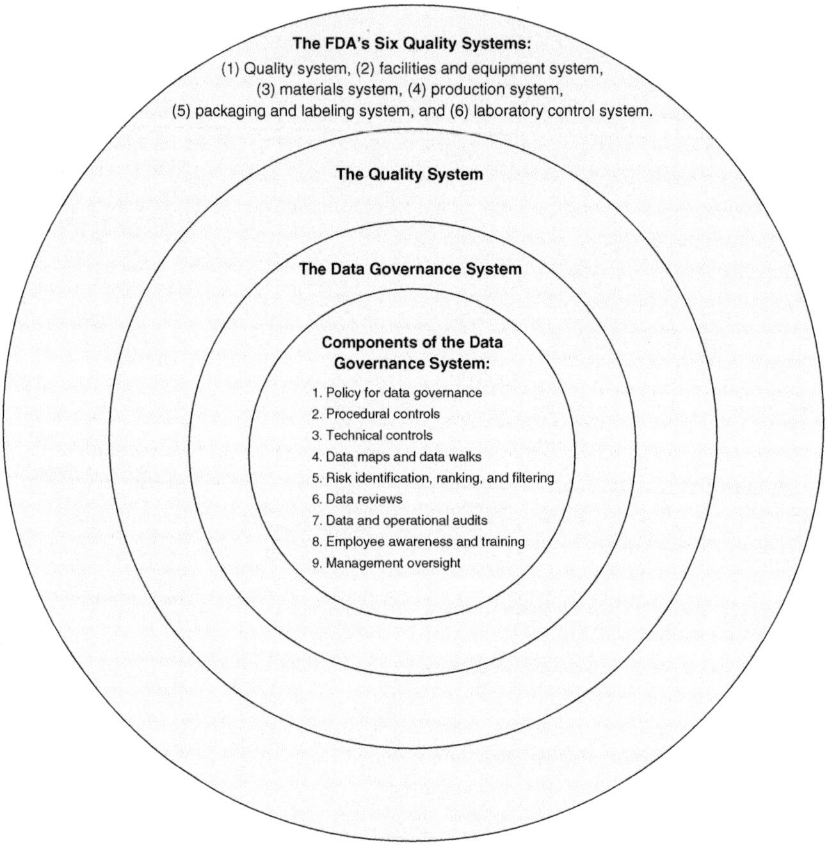

Figure 9.2 Data governance and data integrity hierarchy.

Contents of the Sub Element

The nine components of the data governance system shown earlier are discussed in additional detail in the following sections.

Policy for Data Governance

Data governance is the sum total arrangements, which provide assurance of data integrity. Practically all functioning QC laboratories in the pharmaceutical industry already possess the core elements of a Data Governance system [2, 3]. Often, the only thing they lack is an overarching policy or procedure, which ties all the elements together. This document will be unique to the individual organization, but will typically contain the follow elements.

- An explanation as to the purpose of the Data Governance system.
- A graphical description of data lifecycle (see Figure 9.1).
- A description of roles and responsibilities.
- A statement linking the Data Governance system to the LCS and its sub elements.
- A listing of related SOPs, which constitute the procedural controls component.
- A listing of Data Governance and Data Integrity related terms and their corresponding definitions.
- A short description of the components shown in the Data Governance and Data Integrity hierarchy:
  - Procedural controls
  - Technical controls
  - Data maps and data walks
  - Risk identification, ranking, and filtering
  - Data reviews
  - Data and operations audits
  - Employee awareness and training
  - Management oversight
- Regulatory and industry data integrity references.

Procedural Controls

Procedural controls are SOPs, which provide guidance on subjects that impact directly or indirectly, how data are generated, processed, reviewed, reported, stored, retrieved, archived, and ultimately destroyed. Some typical SOP titles associated with laboratory operations that impact data integrity include the following:

- Laboratory Document Control System
- Laboratory Good Documentation Practices
- Installation, Operational, and Performance Qualification (IQ/OQ/PQ) of Laboratory Equipment
- Laboratory Equipment Lifecycle Management
- Laboratory Building and Facilities Security and Access Controls
- Facility Disaster Recovery Plan
- Analytical Test Method Validation
- Verification of Compendial Procedures
- Electronic Records and Signatures
- Electronic Records Storage, Backup, Archival, and Restoration
- General Procedures for Computer System Validation
- Chromatographic Data Acquisition Software
- Electronic Laboratory Notebooks (ELNs)
- Laboratory Information Management System (LIMS)

- Computer System Change Control Procedures
- Validation of Spreadsheets
- Validation of Databases
- Power Failure Recovery Procedures for Computers
- Disaster Recovery of Electronic Data and Computer Equipment
- Computer System Security
- Operational Maintenance of Computer Systems and Software
- Overview of Data Governance and Data Integrity
- Data Sources and Data Mapping
- Application of Hazard Analysis Critical Control Points (HACCPs) to Laboratory Data Integrity
- Personnel Compliance Program for Insuring Laboratory Data Integrity
- Work Flow, Sample Management, Tracking, Trending, and Release of Analytical Data
- Analytical Data Review and Approval
- Conducting, Documenting, and Reporting Laboratory Investigations, Out-of-Specification (OOS) and Out-of-Trend (OOT) Investigations
- QA Oversight and Monitoring of Production
- QA Oversight and Monitoring of QC Laboratory Operations

Technical Controls

Technical controls are inherent controls within system hardware or software that prevent or restrict users from unauthorized or inadvertent manipulation or deletion of data. Systems utilized for acquisition, processing, and storage of GMP data should have technical controls that are verified through qualification or validation. Where technical controls are deficient, procedural controls, or a combination of procedural and technical controls, will be implemented as required based on risk management principles. Technical controls are always preferable to procedural controls because they exclude or limit a person's ability to manipulate or delete data or records. FDA specifically recommends that "you restrict the ability to alter specifications, process parameters, data, or manufacturing or testing methods by technical means where possible (e.g. by limiting permissions to change settings or data)" [1].

Each organization will possess its own unique set of technical controls. These controls are defined by the type of manufacturing and laboratory equipment and computerized systems, which they employ to make active pharmaceutical ingredients (APIs) and/or drug products.

Data Maps and Data Walks

The most effective means to identify risks to data integrity is to generate process diagrams or work flow diagrams (e.g. Data Maps) of the production and testing steps taken during manufacturing of product. Following the data mapping exercise, it is imperative to verify the accuracy of the generated data maps by performing "data walks": That is manual confirmation of the steps and data sources captured during the data mapping exercise. The best means to accomplish the mapping and walking exercise is through execution of a data integrity risk assessment work plan. Sections of such a plan are presented in Figure 9.3.

Goal:
To implement a systematic risk assessment process to determine the integrity of data in electronic, paper-based, or hybrid formats, in alignment with current industry best practices and current regulatory thinking and guidance.

Objective #1:
To create workflow diagrams (e.g. data maps) via subject matter expert (SME) group discussions of all manufacturing, testing, and distribution efforts with emphasis on identifying data sources and determining associated risk to data integrity.

Note 1: This mapping process will follow the manufacturing and testing workflows beginning with the issuance of the manufacturing order and will end with the release of product for delivery and distribution to customers.

Note 2: More detailed data maps of specific testing and/or instrumentation will also be created for the QC laboratory.

Objective #2:
To verify the accuracy of the generated data maps by performing "data walks"; that is manual confirmation of the steps and data sources captured during the data mapping exercises. This will occur on the manufacturing floor, QC laboratory bench, warehouse floor, and any other pertinent location identified in the data mapping exercise.

Objective #3:
To perform a gap analysis based on processes, data sources, and records identified and reviewed during the data mapping + data walk sessions.

Objective #4:
Perform a data integrity risk assessment on the identified gaps in Objective #3 by using Quality Management tools such as risk ranking and filtering (RRF) techniques.

Approach:
Perform a systematic risk assessment of data integrity for "Your Product A", including both electronic and paper data and records. This will be accomplished by evaluating data flows and methods for generating, recording, processing, using, and retaining data created during the production of "Your Product A" batches. This will cover the full operational scope of the manufacturing process from purchase order issuance to product delivery (see Attachment 9.1: Major Steps in manufacturing, testing, packaging, and release of "Your Product A"). Steps for performing the risk assessment include the following:

- Execute a subject matter expert (SME) team review (e.g. data mapping) of the manufacturing and testing process for each step in the production of "Your Product A" (see Attachment 9.1).
- Capture the review in process diagrams (e.g. data maps).

Figure 9.3 Example data integrity risk assessment work plan.

- Perform a "data walk" to manually confirm the steps and data sources documented in the data maps generated during the data mapping exercises.
- Identify and document gaps to data integrity; Use ALOCA+ as a guide (see Attachment 9.2).
- Perform a gap analysis.
- Perform a data integrity risk assessment by using Quality Management tools such as risk ranking and filtering (RRF).
- Implement interim controls for critical data integrity risks.
- Use data maps and risk assessment tool to design and implement a Quality Assurance Data and Operations Audit Program.

Outcome and Outputs:

- Mapped "Your Product A" workflow (data maps) with emphasis on data sources and associated risk to data integrity from manufacturing order issuance to product delivery.
- On-floor/at-bench/in warehouse, etc. confirmation of the data maps.
- Identification and documentation of gaps associated with data integrity.
- A completed gap analysis.
- Interim controls to address critical data integrity risks.

Timing and Timeline:

To be determined.

Note: Critical data integrity challenges, if identified, will be addressed as they discovered.

Attachments:

1. Attachment 9.1: Major steps in manufacturing, testing, packaging, and release of "Your Product A"
2. Attachment 9.2: Criteria of data integrity: ALCOA+

References:

1. Section 5.3.3 of "Good practices for data management and integrity in regulated GMP/GDP environments" Pharmaceutical Inspection Convention Pharmaceutical Inspection Co-Operation Scheme (PIC/s) PI 041-1 (Draft 2) 10 August 2016.
2. MHRA GxP Data Integrity Definitions and Guidance for Industry Draft, version for consultation July 2016.
3. U.S. Department of Health and Human Services Food and Drug Administration Center for Drug Evaluation and Research (CDER) Center for Biologics Evaluation and Research (CBER) Center for Veterinary Medicine (CVM) (2016). Data integrity and compliance with CGMP guidance for industry. Draft Guidance, April.
4. World Health Organization (2015). Guidance on good data and record management practices, September, *Working document QAS/15.624*.
5. World Health Organization (2003). Application of Hazard Analysis and Critical Control Point (HACCP) methodology to pharmaceuticals, Annex 7. *WHO Technical Report Series No. 908*.
6. ICH HARMONISED TRIPARTITE GUIDELINE (2005). QUALITY RISK MANAGEMENT, Q9, Current Step 4 version, dated 9 November.
7. ICH HARMONISED TRIPARTITE GUIDELINE (2008). PHARMACEUTICAL QUALITY SYSTEM, Q10, Current Step 4 version dated 4 June.

Figure 9.3 (*Continued*)

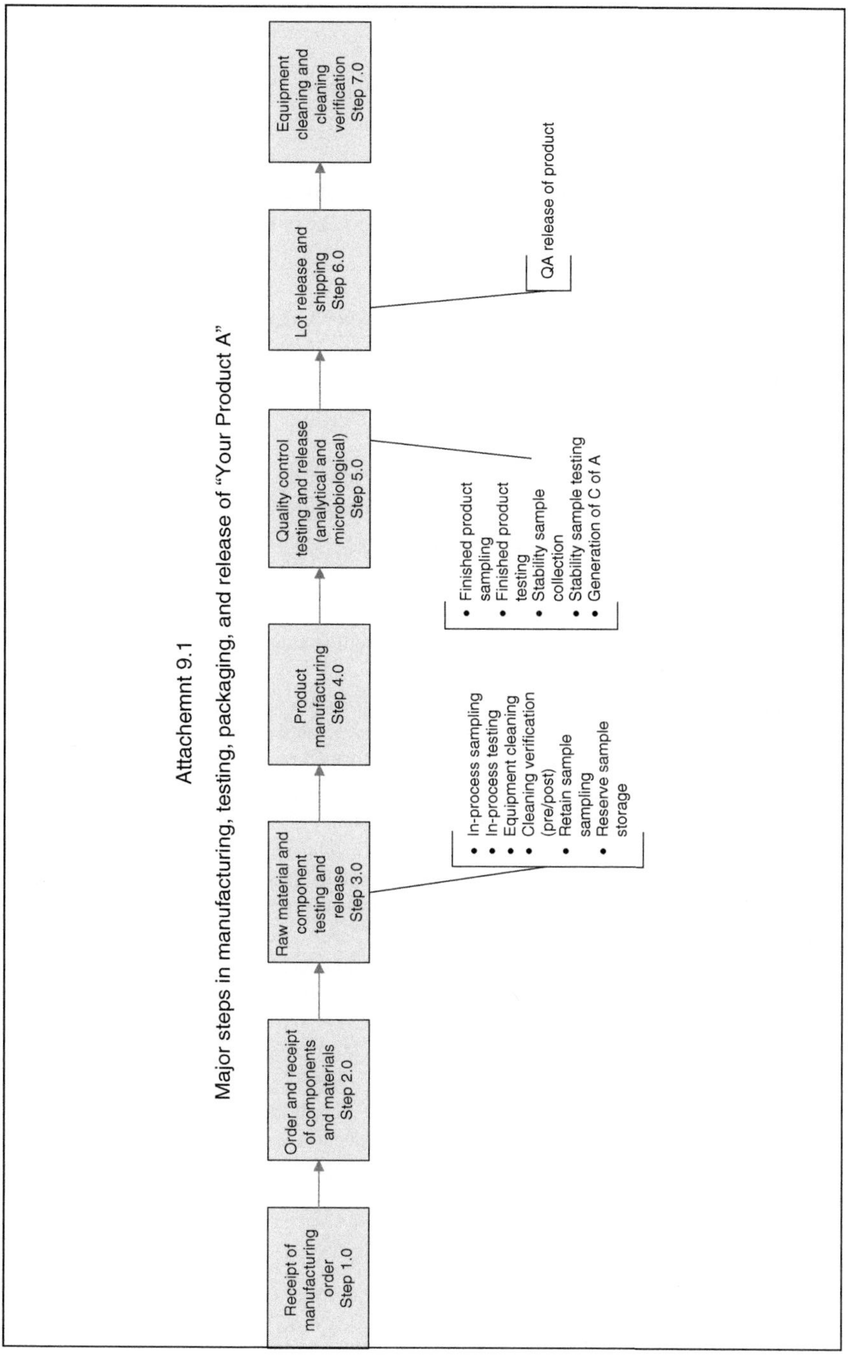

Figure 9.3 (*Continued*)

Attachment 9.2

Criteria of data integrity: ALCOA+

Criterion	Meaning
Attributable	Attribute to the person generating the data (MHRA) Who acquired the data originally or performed an action subsequently to it and when?
Legible	Legible (MHRA) Can you read the data together with any metadata or all written entries on paper?
Contemporaneous	Contemporaneous (MHRA) Documented (on paper or electronically) at the time of an activity
Original	Original record or true copy (MHRA) Written observation or printout or a certified copy thereof Electronic record including metadata of an activity
Accurate	Accurate (MHRA) No errors in the original observation(s) No editing without documented amendments/audit trail entries by authorized personnel
Complete	All data from an analysis, including any data generated before a problem is observed, data generated after repeat part or all of the work or reanalysis performed on the sample For hybrid systems, the paper output must be linked to the underlying electronic records used to produce it
Consistent	All elements of the analysis, such as the sequence of events, follow on and data files are date (all processes) and time (when using a hybrid or electronic systems) stamped in the expected order
Enduring	Recorded on authorized media, e.g. laboratory notebooks, numbered worksheets, for which there is accountability or electronic media Not recorded on the backs of envelopes, laboratory coat sleeves, cigarette packets or post-it notes
Available	The complete collection of records can be accessed or retrieved for review and audit or inspection over the lifetime of the record

Figure 9.3 *(Continued)*

An editable form of this Data Integrity Risk Assessment Work Plan is provided in electronic format in Example Tools and Templates, which can be found in the Chapter 9 Appendix (www.wiley.com/go/Bliesner/LabControl_GMPEnvironment).

The output of the mapping exercise, as described in the example work plan, is a comprehensive data map, which is a process diagram of the major manufacturing, testing, packaging, and release steps. Data maps should be as comprehensive and detailed as possible, to ensure that all risks to data integrity have been identified.

Before attempting your data mapping exercise, make sure you do some research within your organization. There is a good chance that complete process diagrams or at least partial diagrams have been previously generated by either your engineer or process improvement groups.

A completed copy of a model data map is included in the Appendices (www.wiley.com/go/Bliesner/LabControl_GMPEnvironment). However, for the sake of brevity, selected sections (e.g. a snapshot of each major step in the process) of this map are presented in Figure 9.4.

After the data mapping exercise for the major manufacturing, testing, packaging, and release steps, it will be necessary to execute more detailed mapping exercises with the laboratory. Specifically, each QC instrument/testing process should be mapped to illustrate how data is created, modified, reported, and managed. Figure 9.5 shows a generic example of a laboratory instrument data map.

At this point in the mapping exercises, the "data walks" need to be performed. Recall that data walks are manual confirmation of the steps and data sources captured during the data mapping exercise.

The data walks are best performed by persons who are not intimately familiar with the manufacturing, testing, packaging, and release process. They will be less likely to make assumptions and/or gloss over sections of the process maps. It should also be noted that data walks can be used as a training tool to introduce personnel to the production process in general.

Risk Identification, Ranking, and Filtering

In addition to the data maps, output of the data mapping and data walking exercise include (i) gaps analyses and (ii) risk assessment tools.

The gap analyses include documentation of observed gaps with linkage to the steps in the data maps. Table 9.2 is a partial representation of the gap analysis linked with the data map depicted in Figure 9.4 titled "Example data map for the manufacture of a drug product." Table 9.3 is partial representation of the gap analysis linked to the data map depicted in Figure 9.5 titled "A generic laboratory instrument data map."

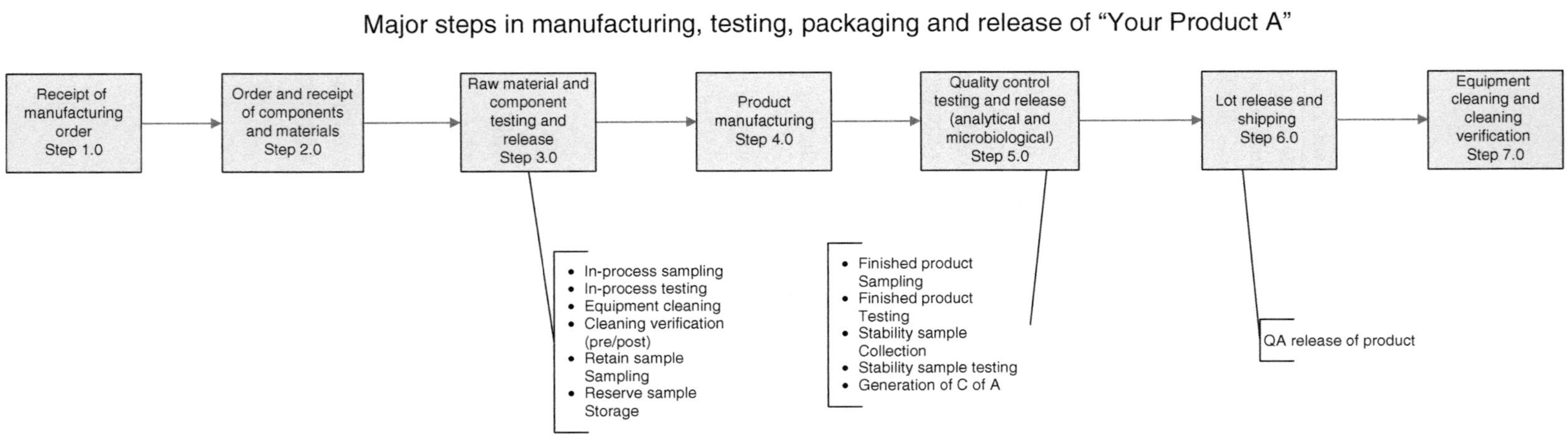

Figure 9.4 Example Data Map for the manufacture of a drug product.

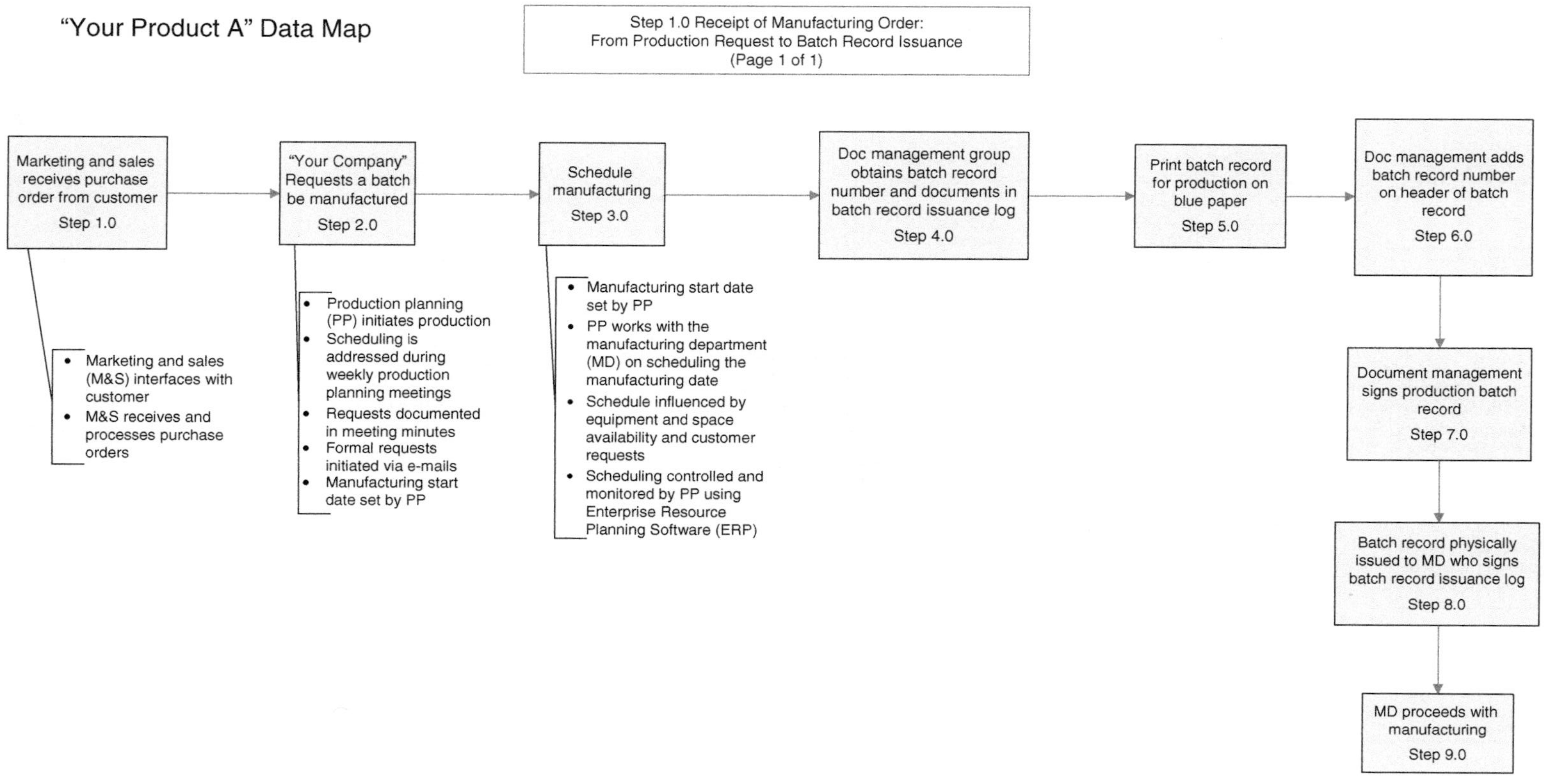

Figure 9.4 (*Continued*)

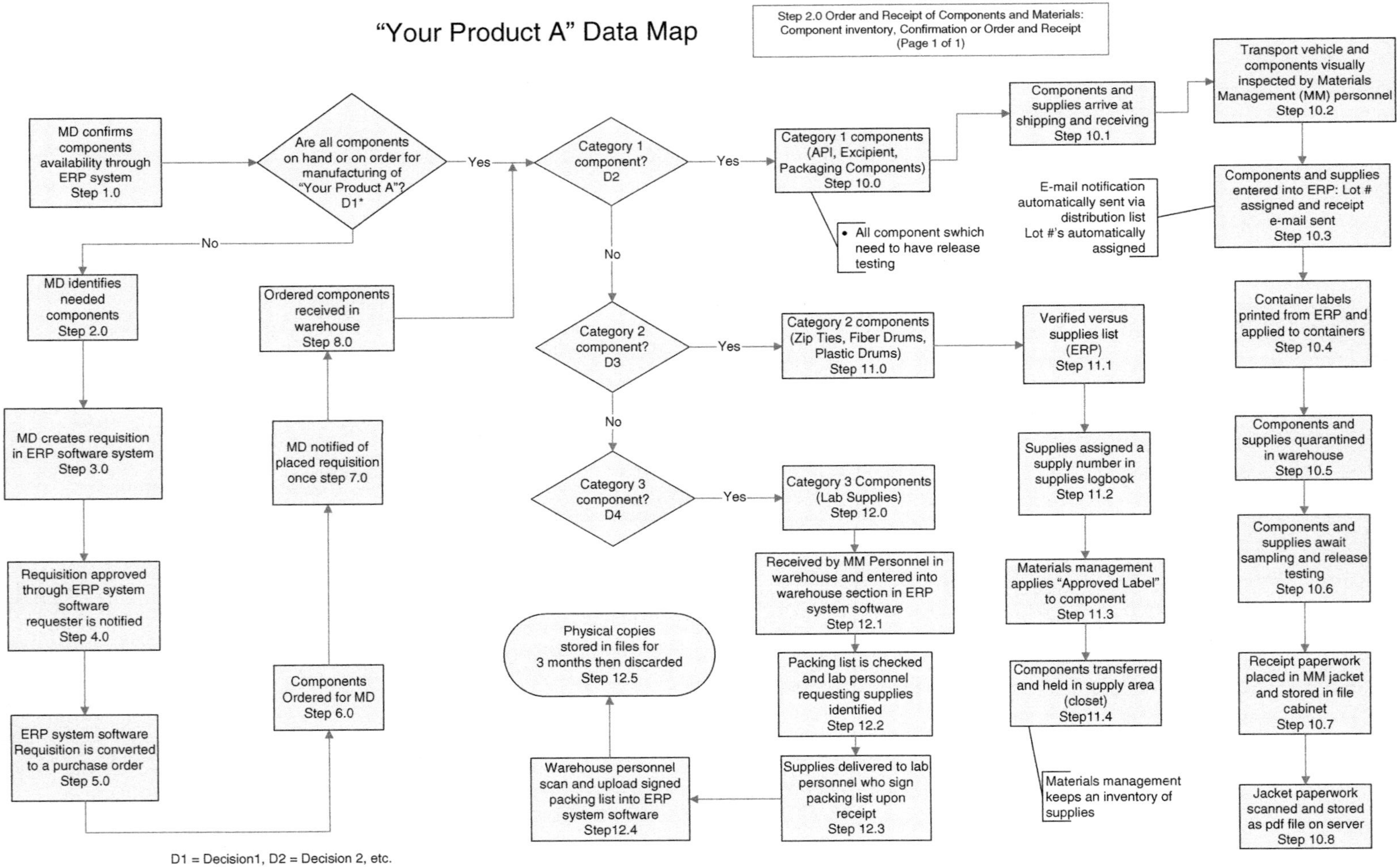

Figure 9.4 (*Continued*)

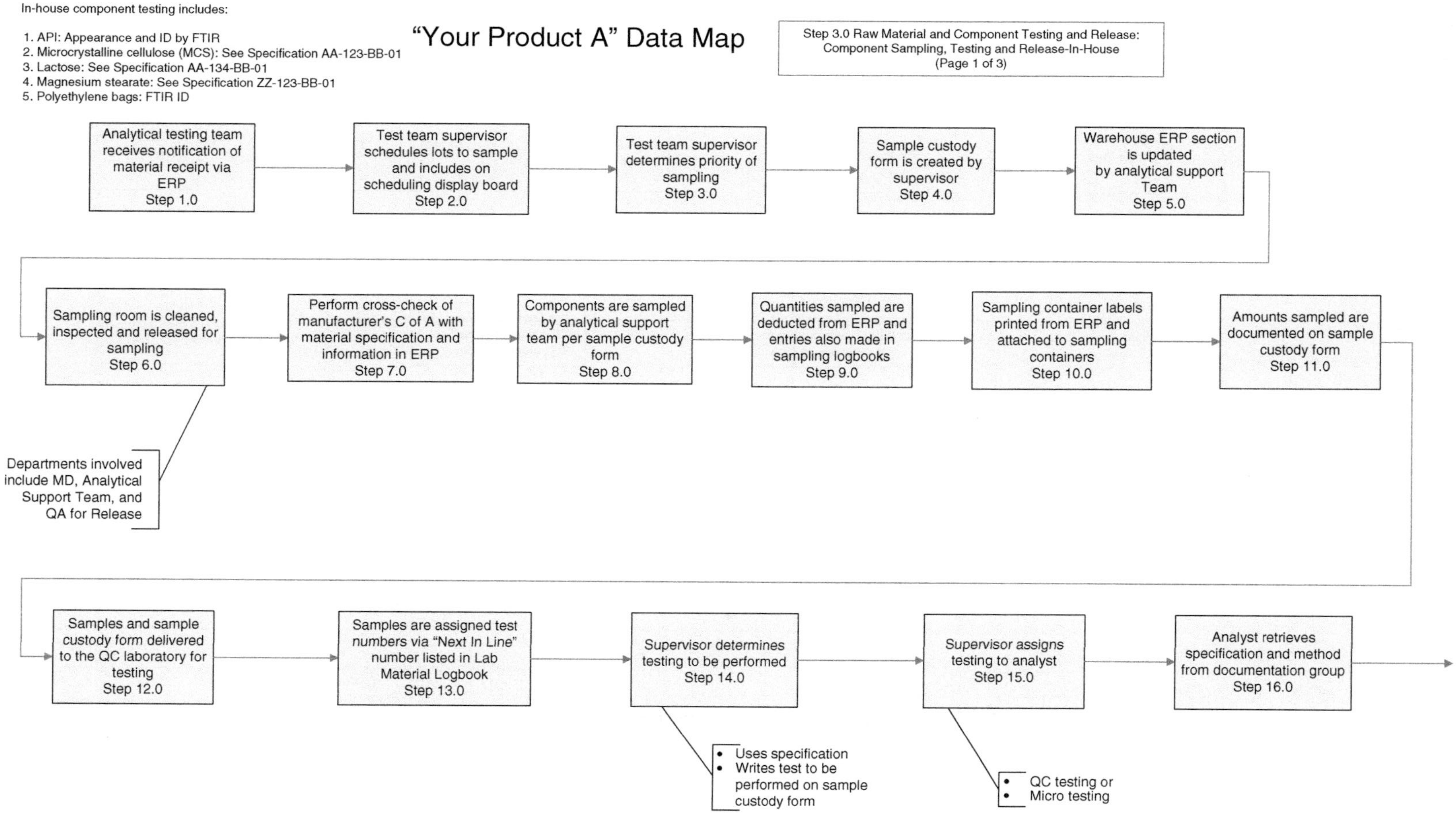

Note: There are a total of 3 pages to this section of the data map. See Chapter 9 Appendix for the additional pages.

Figure 9.4 (*Continued*)

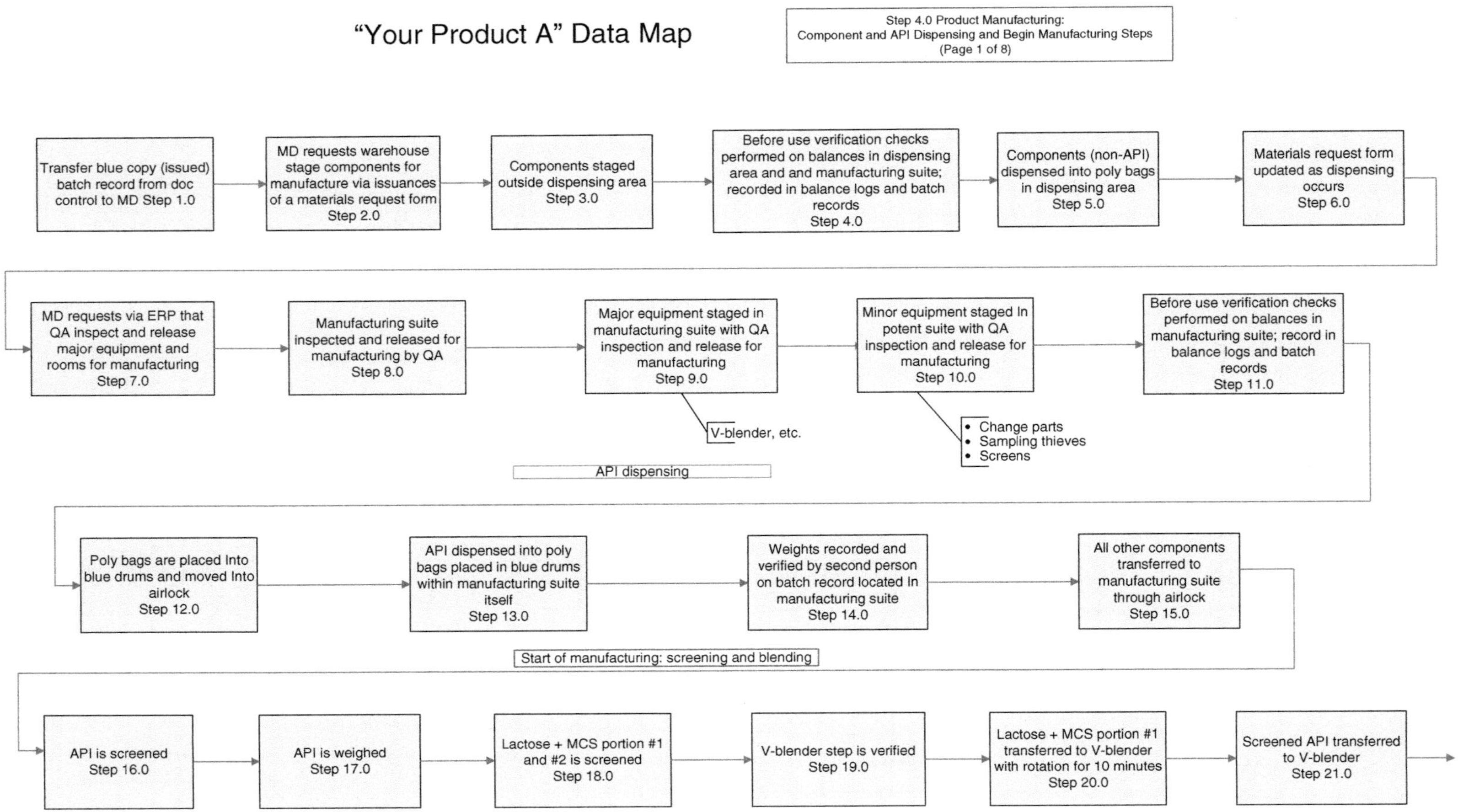

Note: There are a total of 8 pages to this section of the data map. See Appendix for the additional pages.

Figure 9.4 *(Continued)*

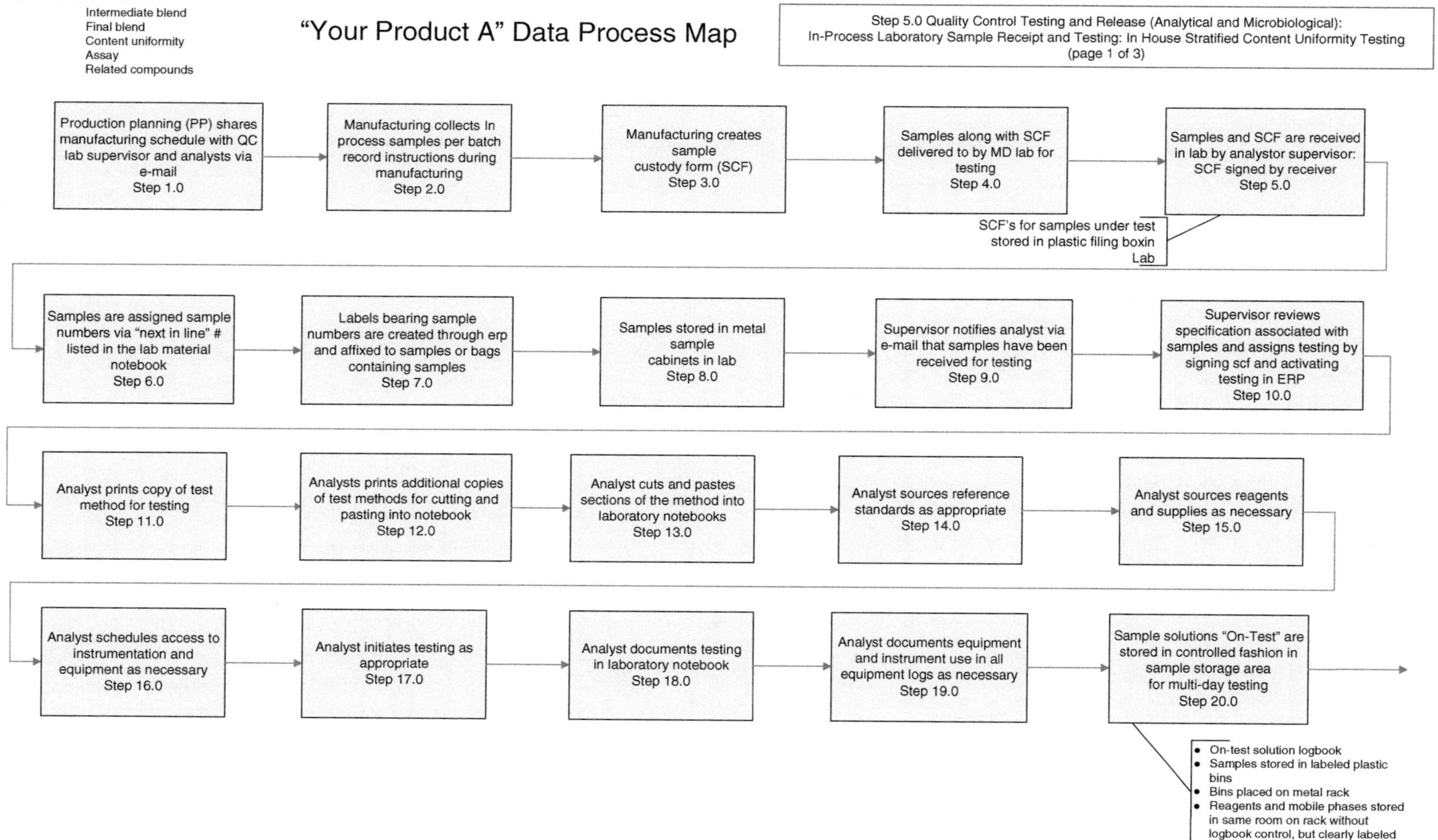

Note: There are a total of 3 pages to this section of the data map. See Appendix for the additional pages.

Figure 9.4 (*Continued*)

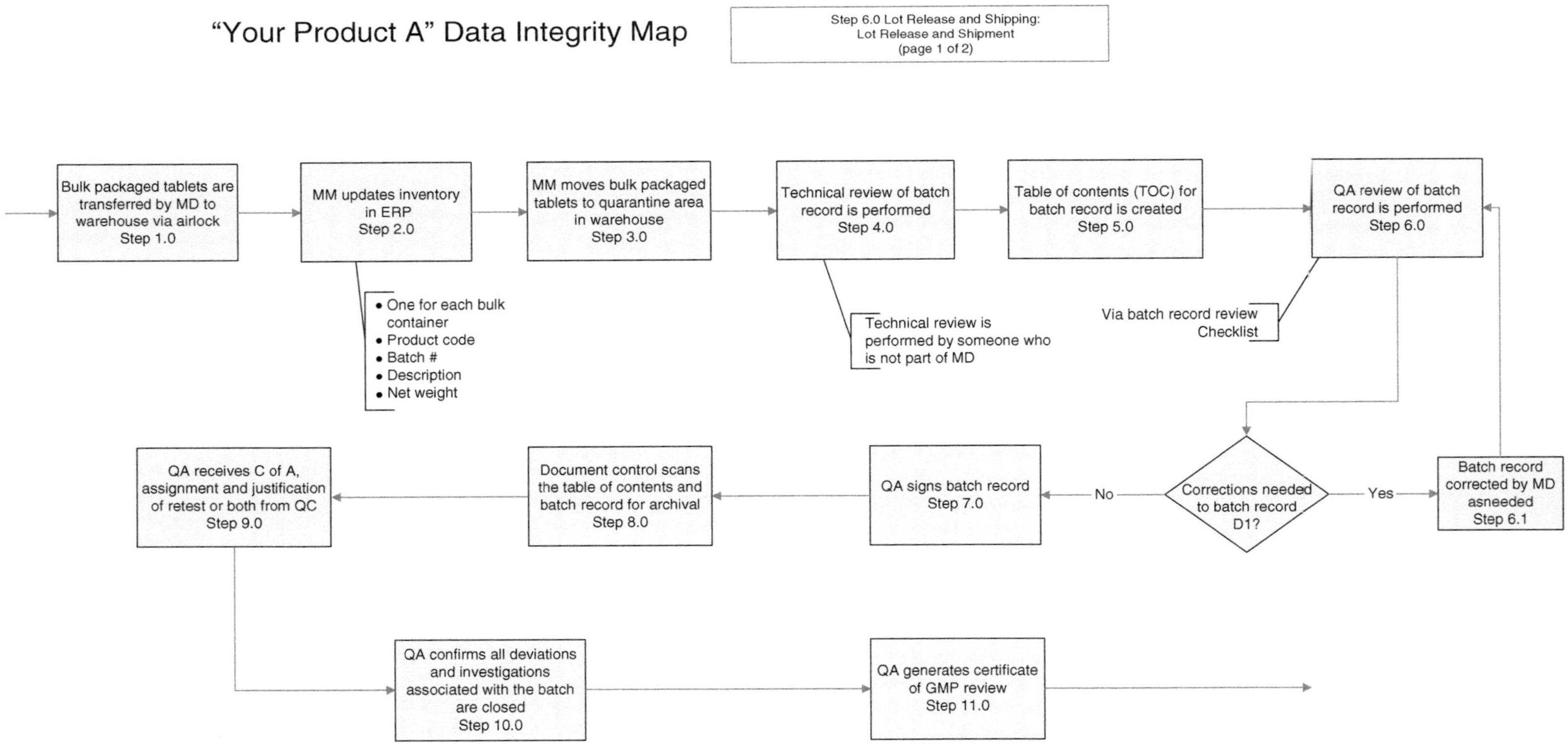

Note: There are a total of 2 pages to this section of the data map. See Appendix for the additional page.

Figure 9.4 (*Continued*)

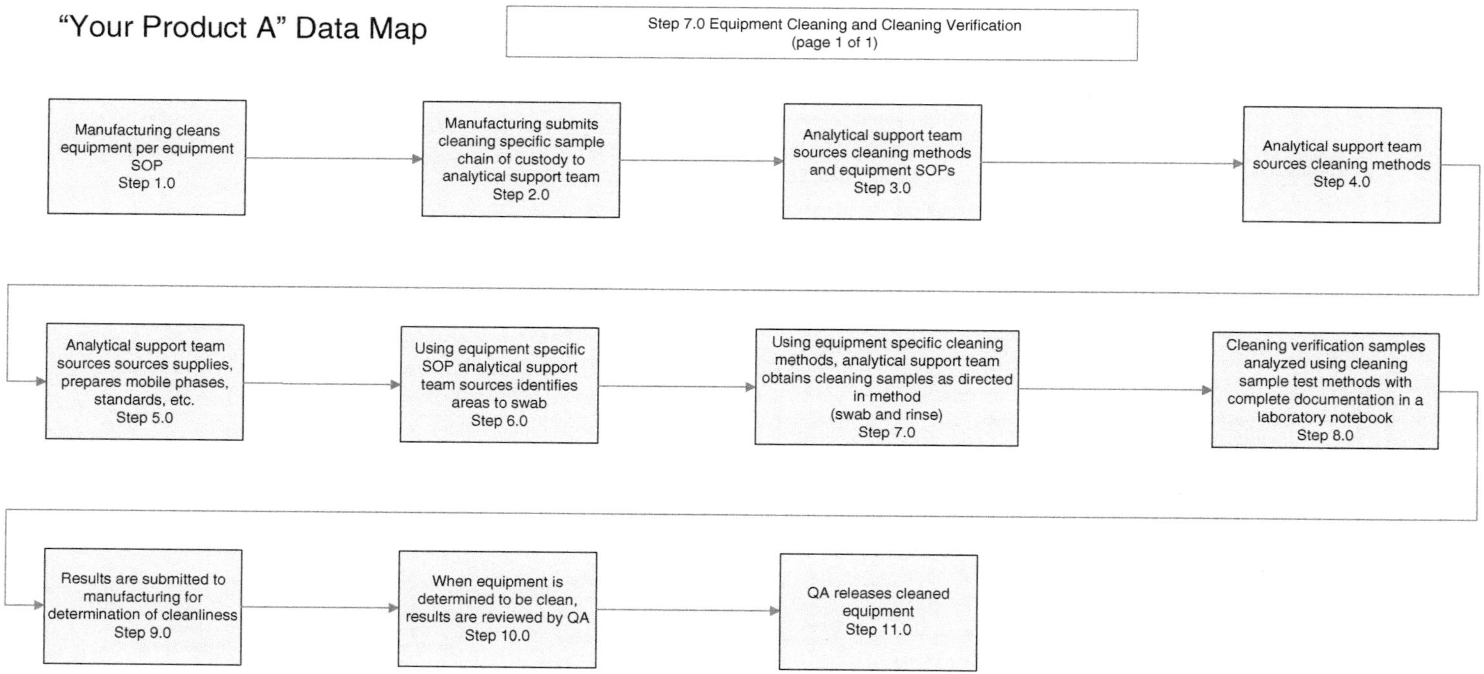

Figure 9.4 (*Continued*)

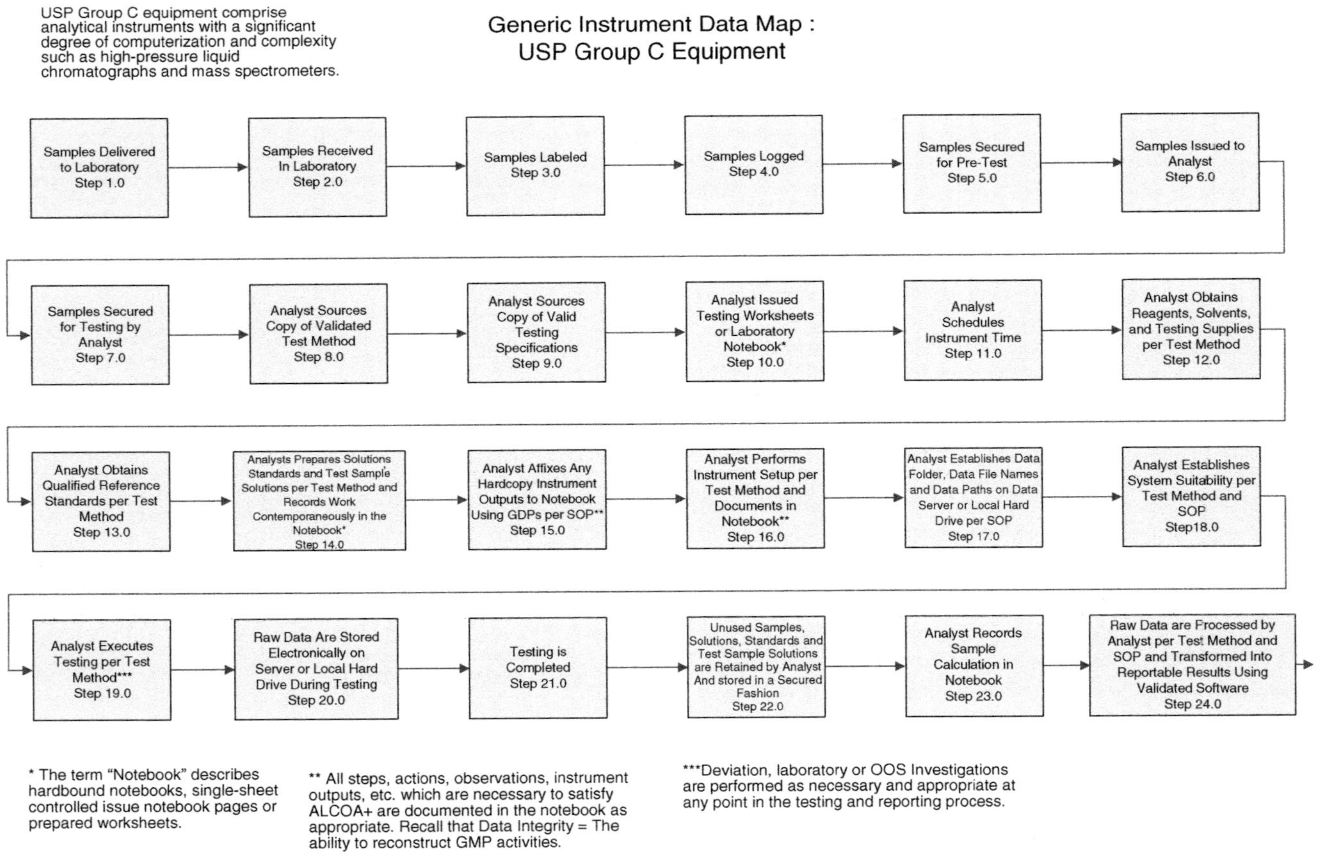

Figure 9.5 A generic laboratory instrument data map.

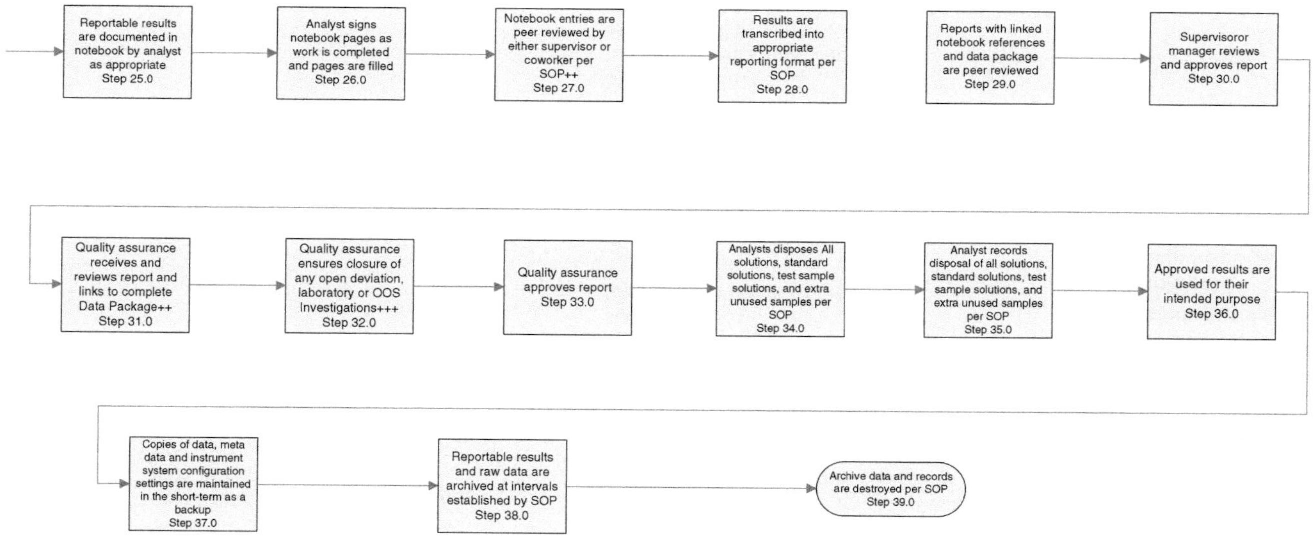

++ Corrections are made as appropriate throughout the review process.

+++ As before, Deviation, Laboratory, or OOS Investigations are performed as necessary and appropriate at any point in the testing and reporting process.

Figure 9.5 (*Continued*)

Table 9.2 Example gap analysis for manufacturing steps.

#	Source step	Observed gap	References	Explanatory notes	ALCOA+ compliance challenges	Suggested corrective and preventive actions
1	General quality system review	Resources do not currently exist to support and sustain a Data Governance System and encourage good data integrity practices.	Your Audit Notebook, p. 157 PIC/S Section 6.6 Resource Allocation WL: 320-13-20 Fresenius Kabi AG	• Observation is supported by the fact that data integrity was first flagged by QA in September 2025 and was not formally addressed until December 2026, with limited progress to this date. • Workload and pressures experienced by those responsible for data generation and record keeping increases the likelihood of errors or the opportunity to unintentionally and/or deliberately compromise data integrity. • An inherent problem and risk which is born by the Contract Manufacturing and Testing industry: multiple clients, high client expectations, unrealistic timelines, with zero defect mentality.	Lack of resources (e.g. Time, equipment, personnel, space, and budget) challenges all aspects of ALCOA+	• Separate commercial manufacturing and testing from development and clinical • Dedicate QA and QC resources to support commercial manufacturing • Design new quality system to support commercial manufacturing • Implement commercial QA quality system

2	5.0 Quality control testing and release review	Systematic analytical data review metrics, as described in SOP-DI-LAB-001 "Analytical Data Review" is not formally captured, tracked and trended.	Your Audit Notebook, p. 157, p. 152 URM p. 171 PIC/S Section 6.4.2 and 6.5 Regular management review of quality metrics	• Other than a second person signature in the laboratory notebook, there is no documented evidence to show second party review of analytical data was performed according to SOP in SOP-DI-LAB-001 "Analytical Data Review" • No checklist is attached to the SOP • Data reviewers create their own make-shift review checklists on notebook paper to assist in review; would be cited by an inspector • Data review can provide sources of metrics for analysis and reporting by QA and management • There should be regular management reviews of quality metrics, including those related to data integrity, so that significant issues are identified, escalated, and addressed in a timely manner.	ALCOA+ Principles = **C**omplete	• Develop existing SOP to make checklists, which are attachments • Document review using written checklist which becomes part of the batch release record. • Use checklists to capture metrics related to data review. • Periodically analyze and report review metrics.

(Continued)

Table 9.2 (Continued)

#	Source step	Observed gap	References	Explanatory notes	ALCOA+ compliance challenges	Suggested corrective and preventive actions
3	2.0 Order and Receipt of Components and Materials Review	GMP data is sometimes recorded on sticky notes.	Your Audit Notebook, p. 152 PIC/S 8.4 Expectations for the generation, distribution, and control of records	Personnel were observed using sticky notes to capture GMP information in the warehouse.	ALCOA+ Principles = **O**riginal **E**nduring	• Root-cause analysis should be performed to determine why this practice exists. • Integrate these unacceptable documentation practices into data integrity training course. • Make this a "must review" item for QA review of on-floor, at-bench, at-desk observation of work.

4	6.0 Lot release and shipping review	Use of uncontrolled and/or unvalidated spreadsheets to make GMP lot release decisions.	Your Audit Notebook, p. 100; Pages 1, 2, 3, and 5 Your Product Data Integrity Process Map See Warning Letter: 320-17-17, FACTA Farmaceutici S.p.A. for official entries into spreadsheet for transfer into official forms at a later time.	• Warehouse Tracker spreadsheet is used to make GMP decisions. • Manufacturing Department (MD) Spreadsheet is used in scheduling and control of manufacturing work flow • Supplies List is used to track inventory of Category 2 components such as zip ties, fiber drums, plastic drums, etc. • Quality Incident Spreadsheet lists any open quality investigations and is used to make GMP lot release decisions. o Spreadsheet needs to be reconciled periodically versus the quality event log, which is the official record of open investigations. o Product with open quality investigations can be mistakenly released	ALCOA+ Principles = **O**riginal	• Perform "what if" brainstorming exercises with SMEs and End Users and QA personnel to assess the possibility for elimination of spreadsheets by asking what impact their removal will have. • If sheets can't be eliminated, then determine a means to control them or find another secure electronic system to replace them. • Investigate any and all means to eliminate the use of spreadsheets altogether.

(Continued)

Table 9.2 (Continued)

#	Source step	Observed gap	References	Explanatory notes	ALCOA+ compliance challenges	Suggested corrective and preventive actions

5	3.0 Raw material and component testing and release	Not all balances used for GMP activities have attached printers.	URM p. 177;
			
			PIC/S Section 8.10 Direct print-outs from electronic systems
			MHRA GxP data integrity definitions and guidance for industry page 4 designing systems to assure data quality and integrity
				• Observed in sampling area
				  o No printers attached to balances
				  o Weighing's in sampling area are not witnessed and confirmed in writing by a second person
				• Observed in QC lab for certain balances.
				  o Not all balances have attached printers
				  o Weighing's in QC lab are not witnessed and confirmed in writing by a second person
				• Expectation for QC labs in particular is that weigh tapes are necessary
					ALCOA+ Principles =
					Original
					Complete
						• Purchase printers for attachment to all analytical balances in the QC laboratory
						• Consider balance printers for operations in the production area including within glove boxes as necessary.

6	4.0 Product manufac-turing	Entries into the production batch record are not always made in a contemporaneous fashion.	URM, p. 180. PIC/S Section 7.5 Basic DI principles applicable to both paper and electronic systems (ALCOA+): WL: 320-14-01 Wockhardt Limited Ltd.	• Observation was made during API dispensing. • Observation was noted to Supervisor and practice changed to contemporaneous documentation	ALCOA+ Principles = **C**ontemporaneous	• Install printers in glove boxes for scales and balances used in manufacturing. • Attach the printed weigh tapes to the batch record.
	Etc.	Etc.	Etc.	Etc.	Etc.	Etc.

Steps = 1.0 Receipt of Manufacturing Order; 2.0 Order and Receipt of Components and Materials; 3.0 Raw Material and Component Testing and Release; 4.0 Product Manufacturing; 5.0 Quality Control Testing and Release; 6.0 Lot Release and Shipping; 7.0 Equipment Cleaning and Cleaning Verification; General Quality System Review = General Quality System observations.
ALCOA+ Principles = **A**ttributable, **L**egible, **C**ontemporaneous, **O**riginal, and **A**ccurate+ **C**omplete, **C**onsistent, **E**nduring, and **A**vailable.

Table 9.3 Example gap analysis for a laboratory instrument data map.

Step	Step-related conditions or events that may create risk to integrity of data and records	Explanation and comments	Suggestions for technical controls	Suggestions for procedural controls
Step 1.0 Samples delivered to laboratory	1. Samples subjected to temperature and humidity conditions which are outside the labeled storage conditions. 2. Samples physically damaged in transport. 3. Samples lost in transport. 4. Samples misplaced upon delivery. 5. Samples diverted in transport. 6. Samples exchanged in transport.	Transport can be within the same site, the same building, or between sister sites and/or contractor sites. Exchange implies purposeful substitution of samples which will give reportable results versus sample which may give failing results. Diversion in this case means theft: Sample that are of value on the black market.	Execute shipping studies to support investigations when temperature and humidity conditions occur during transport. Execute packaging studies to support investigations when physical damage to packaging occurs during transport. Implement Bar code tracking, RFID tracking per current serialization requirements.	Implement a sample handling and transport SOP which requires two-person verification at all points in sample exchange during transport.
Step 2.0 Samples received in laboratory	1. Unrestricted access to laboratory for sample delivery. 2. "Dump and Run" sample delivery. 3. No personnel designated and trained in sample receipt, storage, and management.	Only laboratory or laboratory interfacing personnel should have access to laboratory areas.	Limit access to laboratory via card key, password, or combination of techniques.	Implement an SOP which addresses *Work Flow, Sample Management, Tracking, Trending, and Release of Analytical Data*. Hire or train a person who is responsible for sample receipt, storage, and management.

(Continued)

Table 9.3 (Continued)

Step	Step-related conditions or events that may create risk to integrity of data and records	Explanation and comments	Suggestions for technical controls	Suggestions for procedural controls
Step 3.0 Samples labeled	1. Samples not labeled-accidentally. 2. Samples mislabeled-accidentally. 3. Samples not labeled-purposely. 4. Samples mislabeled-purposely. 5. Sample labels are incomplete. 6. Use of controlled labels.	The level of control expected over unscheduled sample types (e.g. non-opioid) has increased substantially in recent years due to data integrity issues. Uncontrolled labels (e.g. Ones that can be generated within an office software program) are not considered appropriate since labels can be created "at will" and are uncontrolled.	Purchase, install, and qualify a sample tracking system. Activate models within Stability Software package which automatically create sample labels.	Implement an SOP which addresses *Work Flow, Sample Management, Tracking, Trending, and Release of Analytical Data*.
Step 4.0 Samples logged	1. No sample log exists. 2. Samples not logged. 3. Samples incorrectly logged. 4. Sample logs lost or misplaced. 5. Electronic sample logging and tracking system not validated.	A sample log must be maintained to insure the sample chain of custody.	Implement and validate an electronic sample logging and tracking system.	Implement an SOP which delineates the usage of the electronic sample logging and management system.

Step 5.0 Samples secured for pre-test	1) No dedicated "pre-test" storage area. 2) Sample "pre-test" storage area not secure. 3) Samples stored at bench, in drawers or at desks.	Samples which are "pre-test" or "on-test" are frequently left out on the benchtop during testing and data write-up. This is no longer considered an acceptable practice.	Dedicate a secure sample storage area with limited access, managed by a person who is responsible for sample receipt, storage, and management.	Implement an SOP which addresses *Work Flow, Sample Management, Tracking, Trending, and Release of Analytical Data*
Step 6.0 Samples issued to analyst	1) No chain of custody records. 2) Samples analyzed at remote Sites are not analyzed in a timely fashion.	See earlier.	Ensure work flow management software and tools such as ERP have a time-prompt to ensure samples which are sent to remote Sites or Contractors are analyzed in a timely manner.	Implement an SOP which addresses *Work Flow, Sample Management, Tracking, Trending, and Release of Analytical Data* Ensure SOP *Laboratory Administration and Operations,* addresses Quality Agreements with Contract Testing Laboratories have a section which emphasize sample chain of custody management as well as timing and timeline requirements for sample testing and results reporting.
Step 7.0 Samples secured for testing by analyst	1) No area for analyst to secure samples which are "on-test." 2) Existing "on-test" sample storage areas are not secure; open to all personnel.	See earlier	Dedicate a secure sample storage area with limited access, managed by a person who is responsible for sample receipt, storage and management. Include "Pre-Test" areas "On-Test"	Implement an SOP which addresses *Work Flow, Sample Management, Tracking, Trending, and Release of Analytical Data*

(Continued)

Table 9.3 (Continued)

Step	Step-related conditions or events that may create risk to integrity of data and records	Explanation and comments	Suggestions for technical controls	Suggestions for procedural controls
Step 8.0 Analyst sources copy of validated test method	1) No means to confirm validation status of the method. 2) Copies of validated test methods are not current. 3) Test methods can be printed "at will" by analyst with no version control. 4) Printed methods do not have time-stamp expiration of usage dates.	None.	Electronically restrict printing of test methods. If test methods can be printed, ensure printed methods have a time-stamp expiration date of usage imprinted on them.	Maintain controlled copies (electronic or hard copy) of methods validation reports which are accessible to QC analysts. Implement a document control system SOP which addresses printing and control of test methods and testing specifications.
Step 9.0 Analyst sources copy of valid testing specifica-tions.	1) No means to confirm validity of testing specifications. 2) Copies of testing specifications are not current. 3) Testing specifications can be printed "at will" by analyst with no version control. 4) Printed testing specifications do not have time-stamp expiration of usage dates.	None.	Electronically restrict printing of testing specifications. If testing specifications can be printed, ensure printed methods have a time-stamp expiration date of usage imprinted on them.	Implement a document control system SOP which addresses printing and control of test methods and testing specifications.

Steps 10.0 to 39.0	Etc.	Etc.	Etc.	Etc.
Step 39.0 Archive data and records are destroyed per SOP	1) Complete data lifecycle including record destruction is not codified within an SOP. 2) Procedure for data and record destruction are not followed as written. 3) Records and data are destroyed too soon. 4) The process for destroying data and records is not described in detail within an SOP.	A minimum 10 years retention time is recommended, with many organizations tending to keep all GMP data and records for an indefinite period of time.	A designated secure area or facility (e.g. cabinet, room, building, or computerized system) for the long term, retention of data and metadata for the purposes of verification of the process or activity.	Ensure *Data Governance and Data Integrity* SOP addresses data lifecycle management which include data and record destruction. Destruction may be performed by contractors with secure record destruction capabilities.

The complete example gap analysis for a laboratory instrument data map is provided in its entirety in the Chapter 9 Appendix (www.wiley.com/go/Bliesner/LabControl_GMPEnvironment).

Once all the observed gaps have been identified, a risk analysis is performed. This can be accomplished in numerous different fashions. However, Table 9.4 titled "Example risk ranking and filter tool" shows an example which uses the quality tool referred to as risk ranking and filtering (RRF).

Table 9.4a Example risk ranking and filtering tool: Risk definitions.

C1 =	**Data Criticality** = Impact on decision making, product quality and patient safety, namely: (A) Does the data influence important decisions? (B) What is the impact of the data on product quality or patient safety?		
High = 10 Medium = 5 Low = 1 NA = 1			
R1=	**Risk to Data** = Susceptibility to unauthorized: (A) Alteration of data and records (B) Deletion of data and records		
High = 10 Medium = 5 Low = 1 NA = 1			
D1 =	**Detectability** = Likelihood of detection/visibility of changes, alterations or deletion of data and records with existing data integrity procedures and practices		
Easily Detected = 1 Might be Detected = 5 Difficult to Detect = 10 NA = 1			
F1=	**Frequency** = A qualitative (or if data is available, quantitative) sense of how often the observation or practice occurs within the organization over time: (A) Isolated incident (B) Periodically occurs (C) A Reoccurring issue		
Rarely = 1 Occasionally = 5 Frequently = 10 NA = 1			
	High Risk Score =	> 5000	Red
	Medium Risk Score =	Between 1000 and 5000	Yellow
	Low Risk Score =	< 1000	Green

Table 9.4b Example risk ranking and filtering tool: Risk score calculations

OBSERVED GAP		
Resources do not currently exist to support and sustain a Data Governance System and encourage good data integrity practices.		
DATA CRITICALITY (C1)		
High		
Score = 10		
RISK TO DATA (R1)		
High		
Score = 10		
DETECTABILITY (D1)		
Difficult to Detect		
Score = 10		
FREQUENCY (F1)		
Frequently		
Score = 10		
CALCULATED RISK SCORE		
(C1-SCORE) × (R1-SCORE) × (D1-SCORE) × (F1-SCORE) = RISK SCORE		
10*10*10*10 = 10,000 = A high risk score		
High Risk Score = > 5000	Red	
Medium Risk Score = Between 1000 and 5000	Yellow	
Low Risk Score = < 1000	Green	

OBSERVED GAP		
Systematic analytical data review, as described in SOP-DI-LAB-001 "Analytical Data Review" is not formally captured, tracked and trended.		
DATA CRITICALITY (C1)		
Medium		
Score = 5		
RISK TO DATA (R1)		
Medium		
Score = 5		
DETECTABILITY (D1)		
Might be Detected		
Score = 5		
FREQUENCY (F1)		
Frequently		
Score = 10		
CALCULATED RISK SCORE		
(C1-SCORE) × (R1-SCORE) × (D1-SCORE) × (F1-SCORE) = RISK SCORE		
5*5*5*10 = 1250 = A medium risk score		
High Risk Score = > 5000	Red	
Medium Risk Score = Between 1000 and 5000	Yellow	
Low Risk Score = < 1000	Green	

(continued)

Table 9.4b Some Example Lines from the Risk Ranking and Filtering Tool

#	Source Step	Observed Gap	References	Explanatory Notes	ALCOA + Compliance Challenges	Suggested Corrective and Preventive Actions	Data Criticality (C1)	C1-Score	Risk to Data (R1)	R1-SCORE	Detectability (D1)	D1-Score	Frequency (F1)	F1-Score	Risk Score
1	General Quality System Review	Resources do not currently exist to support and sustain a Data Governance System and encourage good data integrity practices.	Your Audit Notebook p. 157** PIC/S Section 6.6 Resource Allocation WL: 320-13-20 Fresenius Kabi AG ** DMB refers to location of Your Product auditor's personal notebook where observations are documented.	• Observation is supported by the fact that data integrity was first flagged by QA in September 2025 and was not formally addressed until December 2026, with limited progress to this date. • Workload and pressures experienced by those responsible for data generation and record keeping increases the likelihood of errors or the opportunity to unintentionally and/or deliberately compromise data integrity. • An inherent problem and risk which is borne by the Contract Manufacturing and Testing industry: Multiple clients, high Client expectations, unrealistic timelines, with zero defect mentality.	Lack of resources (e.g. Time, equipment, personnel, space and budget) challenges all aspects of ALCOA +	• Separate Commercial Manufacturing and Testing from Development and Clinical • Dedicate QA and QC resources to support commercial manufacturing • Design new Quality System to support Commercial Manufacturing • Implement Commercial QA Quality System	High	10	High	10	Difficult to Detect	10	Frequently	10	10000
2	5.0 Quality Control Testing and Release Review	Systematic analytical data review, as described in SOP-DI-LAB-001 "Analytical Data Review" is not formally captured, tracked and trended.	Your Audit Notebook p. 157, p. 152 URM p. 171, PIC/S Section 6.4.2 and 6.5 Regular management review of quality metrics	• Other than a second person signature in the laboratory notebook, there is no documented evidence to show second party review of analytical data was performed according to SOP in SOP-DI-LAB-001 "Analytical Data Review". • No checklist is attached to the SOP • Data reviewers create their own make-shift review checklists on notebook paper to assist in review; would be challenged by an inspector • Data review can provide sources of metrics for analysis and reporting by QA and Management • There should be regular management reviews of quality metrics, including those related to data integrity, such that significant issues are identified, escalated and addressed in a timely manner.	ALCOA + Principles = Complete	• Develop existing SOP to make checklists which are attachments • Document review using written checklist which becomes part of the batch release record. • Use checklists to capture metrics related to data review. • Periodically analyze and report review metrics to management.	Medium	5	Medium	5	Might be Detected	5	Frequently	10	1250
3	2.0 Order and Receipt of Compo-nents and Materials Review	GMP data is sometimes recorded on sticky notes.	Your Audit Notebook p. 152 PIC/S 8.4 Expectations for the generation, distribution and control of records	Personnel were observed using sticky notes to capture GMP information in the warehouse.	ALCOA + Principles = Original Enduring Contemporaneous	• Root-Cause analysis should be performed to determine why this practices exists. • Integrate these unacceptable documentation practices into data integrity training course. • Make this a "must review" item for QA review of on-floor, at-bench, at-desk observation of work.	High	10	High	10	Might be Detected	5	Occasionally	5	2500

(continued)

Table 9.4b (Continued)

#	Source Step	Observed Gap	References	Explanatory Notes	ALCOA + Compliance Challenges	Suggested Corrective and Preventive Actions	Data Criticality (C1)	C1-Score	Risk to Data (R1)	R1-SCORE	Detectability (D1)	D1-Score	Frequency (F1)	F1-Score	Risk Score
4	6.0 Lot Release and Shipping Review	Use of uncontrolled and/or validated spreadsheets to make GMP lot release decisions.	Your Audit Notebooks p. 100; Pages 1, 2, 3 and 5 Your Product Data Integrity Process Map SeeWarning Letter: 320-17-17, FACTA Farmaceutici S.p.A. for official entries into spreadsheet for transfer into official forms at a later time.	• Warehouse Tracker spreadsheet is used to make GMP decisions • MD (Manufacturing Department) Spreadsheet is used in scheduled and control of manufacturing workflow • Supplies List is used to track inventory of Category 2 Components such as zip ties, fiber drums, plastic drums, etc.) • Quality Incident Spreadsheet, lists any open quality investigations and is used to make GMP lot release decisions. o Spreadsheet needs to be reconciled periodically versus the Quality Event Log, which is the official record of open investigations. o Product with open quality investigations can bemistakenly released	ALCOA + Principles = Original- information is transcribed or read from the spreadsheet which is the original record but not the official record.	• Perform "what if" brainstorming exercises with End Users and QA personnel to assess the possibility for elimination of spreadsheets by asking what impact its removal will have. • If sheet can't be eliminated, then determine a means to control or another electronic system to replace it. • Investigate any and all means to eliminate the use of spreadsheets altogether.	High	10	High	10	Easily Detected	1	Frequently	10	1000
5	3.0 Raw Material and Component Testing and Release	Not all balances used for GMP activities have attached printers.	URM p. 177; PIC/S Section 8.10 Direct print-outs from electronic systems MHRA GxP Data Integrity Definitions and Guidance for Industry page 4 Designing systems to assure data quality and integrity	• Observed in Sampling Area o No printers attached to balances o Weighing's in sampling area are not witnessed and confirmed in writing by a second person • Observed in QC lab for certain balances. o Not all balances have attached printers o Weighing's in QC lab are not witnessed and confirmed in writing by a second person • Expectation for QC labs in particular is that weigh tapes are necessary	ALCOA + Principles = Original Complete	• IPurchase printers for attachment to all analytical balances in the QC laboratory • Consider balance printers for operations in the production area including within glove boxes as necessary.	High	10	High	10	Easily Detected	1	Occasionally	5	500
6	4.0 Product Manufac-turing	Entries into the production batch record are not always made in a contemporaneous fashion.	Your Audit Notebook p. 180. PIC/S Section 7.5 Basic DI principles applicable to both paper and electronic systems (ALCOA +): PIC/S 8.8 Verification of records (secondary checks) WL: 320-14-01 Wockhardt Limited WL: 320-15-17 Unimark Remedies Ltd.	• Observation was made during API dispensing. NOTE: Observation was made to Supervisor and practice changed to contemporaneous documentation.	ALCOA + Principles = Contemporaneous	• Install printers in glove boxes for scales and balances used in manufacturing. • Attached the printed weigh tapes to the batch record.	High	10	High	10	Difficult to Detect	10	Occasionally	5	5000

The complete risk ranking and filtering tool in excel spreadsheet is included in the Chapter 9 Appendix (www.wiley.com/go/Bliesner/LabControl_GMPEnvironment).

The RRF tool can then be used to construct a data and operational audit program, which will be described further in Section "Data and Operational Audits".

Data Reviews

Data review is standard component of any CGMP laboratory. Data review needs to be driven by SOP and executed in a logical and controlled manner. There are often three levels of review: (i) at the bench level, (ii) at the supervisor or management level, and (iii) by quality assurance (QA). Figure 9.6 describes how data typically flows through the laboratory, including when and how it is reviewed. Table 9.5 is an example notebook and data review checklist, which provides the structure for data review.

Data and Operational Audits

As mentioned earlier, once the data mapping, data walk, gap analysis, and risk assessment have been performed, a long-term QA-led data and operational audit program can be implemented. This program builds on the risk assessment, and involves several components. These include:

- Identifying critical oversight points for quality assurance (QA-COP), on the data maps using the RRF tool.
- Determining a level of QA oversight required, based on the organizations overall self-assessment of level of CGMP compliance, such as:
  - *Level 1*. Routine Compliance Challenges: "In Control" and "Stay in Control" operations
  - *Level 2*. Minor Compliance Challenges: Form 483 Observation response level
  - *Level 3*. Major Compliance Challenges: Warning Letter or Substantial (or repeated) Form 483 response level
  - *Level 4*. Critical Compliance Challenges: Potential or self-imposed CD response level
  - *Level 5*. "Out of Control" Compliance Challenges: Mandated CD response level
- Determining the type of oversight required as each QC-COP, such as:
  - **Sample and Review** data, documentation, reports, and results generated by execution of work steps or tasks within mapped processes (Outputs).
  - **Observe** execution of the individual work steps or tasks (Inputs) within the mapped processes (e.g. watch testing, data recording, etc.).
  - **Interview** personnel who perform the individual work steps or tasks within mapped processes (e.g. using identified trends or patterns, solicit feedback from bench chemist, technicians, etc. for their opinion regarding challenges or possible changes or upgrades to methods, processes, documentation, etc.).

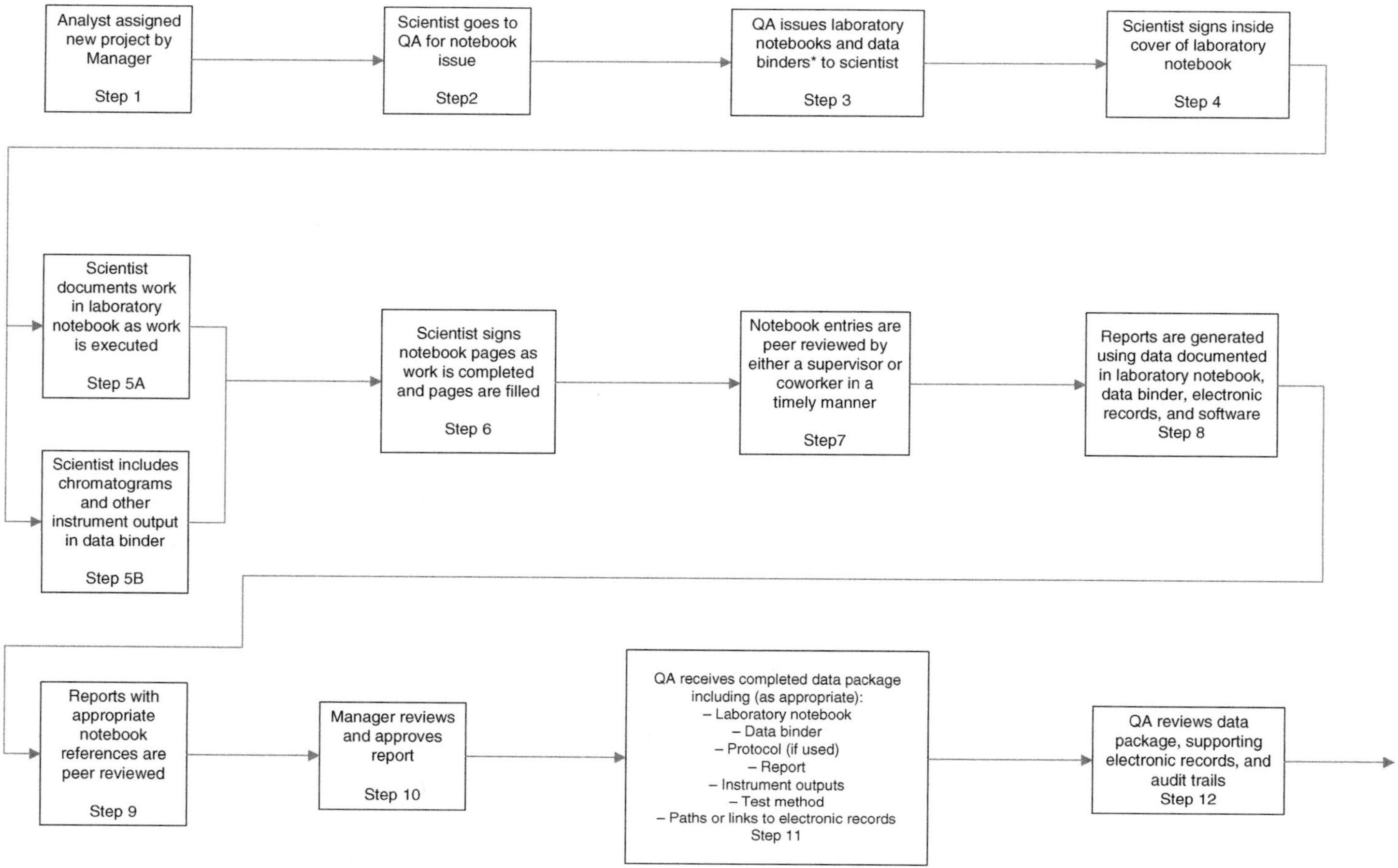

*For this example "data binder" means instrument outputs bound together and controlled.

Figure 9.6 Example laboratory data flow for notebook and data review.

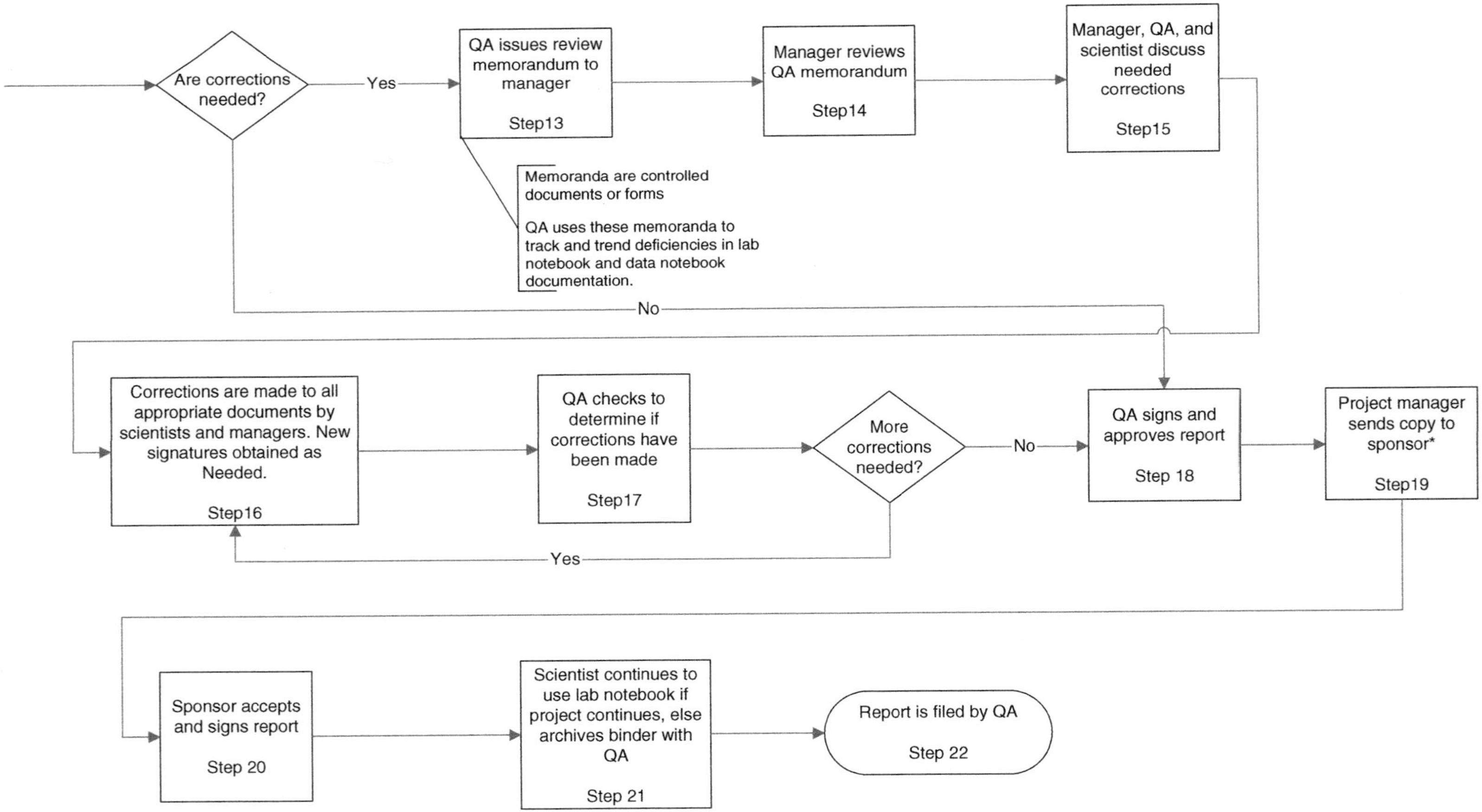

* Sponsor can mean internal or external customers.

Figure 9.6 (*Continued*)

Table 9.5 Example notebook and data review checklist.

			Example notebook and data review checklist
1			**General Requirements**
			Permanently Bound Notebooks
	1.1	☐	Is there a Document Control Manager (DCM) or designee who is responsible for ordering new, bound laboratory notebooks?
	1.2	☐	Are bound notebooks of a defined quality and format and purchased from an approved vendor which insures the following?
			Notebooks are uniquely numbered (pre-numbered if possible)?
			Notebook pages uniquely numbered (pre-numbered if possible)?
			Notebook contains an index?
			Each page has a designated space for all required signatures?
	1.3	☐	When bound notebooks are received, does the DCM or designee, inventory, record receipt, sequentially number, and lock notebooks in a secure area?
	1.4	☐	Does the DCM or designee insure that each issued notebook is free from defects (e.g. no skipped, mis-numbered pages, etc.) prior to issuing the book?
	1.5	☐	Does the DCM or designee record the issuance of each notebook including notebook number, and employee issued to?
	1.6	☐	Is there a limit to the number of active notebooks which can be held by each employee? If yes, is this policy enforced?
	1.7	☐	Does the employee receiving the notebook not only sign the notebook out from the DCM but sign the inside cover of the notebook as well?
	1.8	☐	Are there any restrictions on taking notebooks offsite, and if so, is the policy enforced?
	1.9	☐	Is there any requirement to store notebooks in fireproof cabinets on a daily basis or when not being used by the scientists?
	1.10	☐	Is there a procedure in place to collect notebooks when the employee fills a notebook and needs a new notebook, leaves the company, or no longer needs a notebook?
			Loose-Leaf Binders or Notebooks
	1.11	☐	Is there a procedure in place which allows for the use of loose-leaf binders for the insertion of specimen chromatograms, spectra, or summary print outs from computers?

Table 9.5 (Continued)

	1.12	☐	Are these binders treated as bound laboratory notebooks with respect to data entry, correction, review and permanent storage?
	1.13	☐	Are these binders clearly labeled to identify notebook references, linkages and projects encompassed within?
	1.14	☐	Are scientists and reviewer signatures shown on a cover sheet or label?
			<u>Individual Issue Loose-Leaf Notebook Pages or Worksheets</u>
	1.15	☐	If loose-leaf individual issue notebook pages or worksheets are used, are they issued, controlled, and used in the same manner as permanently bound notebooks?
2			**Data Entry, Documentation, and Corrections**
	2.1	☐	Is there a procedure in place, which clearly states that laboratory data and observations must be recorded in permanently bound uniquely numbered notebooks or on controlled official sheets or forms?
	2.2	☐	Does the procedure require that manually captured data and observations be recorded at the time of their collection or as they are observed?
	2.3	☐	Does the procedure prohibit back dating of notebook entries?
	2.4	☐	Does the procedure specifically require use of indelible ink? (Note: If color is specified is it to be used universally?)
	2.5	☐	Are all notebook entries neat and legible?
	2.6	☐	Are all notebook entries appropriately cross-referenced to corresponding laboratory documentation such as protocols, chromatograms, spectra, etc.?
	2.7	☐	Are the following types of laboratory related documentation linked back to laboratory notebooks as appropriate?
			Methods validation reports?
			Analytical qualification reports?
			IQ/OQ/PQ and calibration records?
			Development and summary reports?
	2.8	☐	Is there a procedure which defines multiple investigator or scientist use of a single notebook?
	2.9	☐	Is there a procedure which defines very clearly how errors should be corrected on a laboratory form or notebook?

(Continued)

Table 9.5 (Continued)

	2.10	☐	Do the instructions for correcting mistakes include:
			Single line out?
			Making the appropriate correction?
			Sign, date and brief description of the correction?
			Provisions for identifying multiple corrections via use of a symbol?
			Use of error codes?
	2.11	☐	If error codes are used, are they consistent within the organization and in compliance with the SOP?
	2.12	☐	Is there a procedure which defines a standard date and time format?
	2.13	☐	Is there a procedure which defines how to secure chromatograms, pictures, written plans, spectra, memos, etc. into the notebook?
	2.14	☐	Is there special considerations given to the handling of thermal paper which fades over time?
	2.15	☐	Is there a specific statement in a procedure which prohibits the use of scratch paper for recording CGMP data, sample calculations, recording observations, etc.?
	2.16	☐	Is there a procedure in place which instruct scientists to avoid leaving blank pages and defines a procedure for crossing out blank or partially blank pages?
3			**Types of Notebook Entries**
	3.1	☐	Is there a standard notebook reporting format which is used by all scientists to aid QA and external data review?
	3.2	☐	Are observations, thoughts, impressions and general notes recorded in the notebook as necessary?
	3.3	☐	Are any and all deviations from written procedures approved, explained, and justified with supervisor and QA approval?
	3.4	☐	Are all laboratory investigations and out of specification results formally and thoroughly investigated and documented?
	3.5	☐	The following types of notebook entries should be made:
			Title of project name, number, etc.

Table 9.5 (Continued)

Test method title and number, including revision	
Protocol number and title	
Brief purpose/objective statement for conducting work	
Brief description/outline of the work to be performed	
Sample identification such as:	
	sample name, codes or ID number
	date of sampling
	date of sample receipt
	quantity of samples received
	who samples were received from
	lot, in-process sample, etc.
	container closure
	stability time point and storage condition
Standard identification such as:	
	name
	source (supplier)
	purity
	lot
	moisture content
	drying procedures used
	expiration date
Reagent identification such as:	
	name
	source
	grade
	lot
	expiration date
	vendor
Sample, standard, mobile phase and reagent preparation steps such as:	
	daily balance calibration checks
	sample weights: manual recordings or balance tape printouts
	pH measurements of solutions

(Continued)

Table 9.5 (Continued)

	size, type and quantity of volumetric glassware	
	dilution schemes	
	mixing times: start and stop	
	sample grind times: start and stop	
	sonication times: start and stop	
	temperatures	
	solution expiration dates	
	sample, standard, mobile phase and reagent solutions identified (labeled)	
Instrument and conditions for all instruments such as:		
	instrument type	
	instrument ID number or similar identifier	
	calibration dates	
	calibration due dates	
	instrument settings and conditions: manual or printout	
	representative instrument printout such as chromatograms and spectra including peak names	
	system suitability acceptance criteria	
	system suitability calculations and proof criteria were met	
	means to capture and document system suitability failures for tracking and trending	
For HPLCs		For GCs
column type, dimensions, and serial number		column type, dimensions, and serial number
detector type and parameters including wavelengths		carrier gas type, make-up flow rates, pressure control settings
mobile phase composition and pH		type of detector
flow rate		temperature of injector and detector
column temperature		note the types and pressures of gases used each day
guard column type		injection volume
injection volume		column temperature program
gradient profile program		run time
injector needle rinse solutions		example injection queue: blanks, standards and samples
run time		computer data storage path name and file names
dwell time		meta data such as integration parameters
example injection queue: blanks, standards and samples		example chromatograms with all peaks labeled
computer data storage path name and file names		investigation of unidentified peaks
meta data such as integration parameters		Standardized method, sequence, processing methods, etc.
example chromatograms with all peaks labeled		Evidence that data processing was conducted per SOP

Table 9.5 (Continued)

			investigation of unidentified peaks	Evidence of stable baselines
			Standardized method, sequence, processing methods, etc.	Headspace setup and operational parameters (as appropriate)
			Evidence that data processing was conducted per SOP	Checks for late eluting peaks
			Evidence of stable baselines	
			Checks for late eluting peaks	
	3.6	☐	Are all manual calculations recorded and shown in detail including proper units of measure, multipliers, dividers, rounding procedures and significant figures?	
	3.7	☐	Is there at least one manual sample calculation for results calculated by computer software?	
	3.8	☐	Are statements made that the conformance to appropriate written specifications for the acceptance of all tested products and materials has occurred? (e.g. a written "pass" or "fail" statement)	
	3.9	☐	Are notebook entries made in sufficient detail as to have the experiments reproduced by a second scientist at a later date without assistance from the original investigator?	
4			**Review and Verification of Data**	
	4.1	☐	Are the notebook pages signed and dated by the individual(s) performing the activity on the date which they occurred?	
	4.2	☐	Are each notebook page signed and dated by a reviewer within an appropriate span of time?	
	4.3	☐	Is the reviewer familiar with the technology to which the experiments relate and should be capable of understanding the significance of the recorded experiments?	
	4.4	☐	Is there evidence that the reviewer is fully reviewing the data entered into the notebook including checking manual calculations and spreadsheet calculations?	
	4.5	☐	Are all data transcriptions checked by the reviewer including data presented in final reports?	
	4.6	☐	Does the reviewer understand that by countersigning the notebook page that they are stating that they have read and understood the entries on that page?	
	4.7	☐	Are all blanks in worksheets and forms filled in with data, observations, or "n/a" or lined out?	

(Continued)

Table 9.5 (Continued)

	4.8	☐	Is there any evidence of "trial injections" of samples?
	4.9	☐	Is there any evidence that analyses have been aborted without explanation?
	4.10	☐	Does a spot check of chromatograms confirm good chromatography and proper integration?
5			**Review of Electronic Laboratory Data and Computerized Systems**
	5.1	☐	Are systems which acquire electronic data properly validated?
	5.2	☐	Are references to the location of electronic data entered into to the notebook (e.g. data path)?
	5.3	☐	Are electronic data stored in folders with logical location names such as finished product testing, active ingredient testing, stability testing, etc.?
	5.4	☐	Are example instrument outputs and meta data printed and affixed in the notebook?
	5.5	☐	Are electronic records controlled with the same care as paper records?
	5.6	☐	Are electronic records backed up on a regular basis?
	5.7	☐	Are electronic records archived on a regular basis?
	5.8	☐	Can electronic records be retrieved from archival within a reasonable period of time?
	5.9	☐	Is review, re-integration, re-processing and other types of modification of electronic data captured in an audit trail?
	5.10	☐	Are audit trails periodically and systematically reviewed?
	5.11	☐	Is access to computerized system controlled by unique usernames and passwords?
	5.12	☐	Do passwords expire and have to be periodically reset?
	5.13	☐	Is access to computerized systems restricted by implementation of a CRUD (Copy, Read, Update, and Delete) matrix or similar document?
	5.14	☐	Do computerized systems have system time-outs which lock entry after a period of inactivity?
	5.15	☐	Is there any evidence that passwords are shared?
	5.16	☐	Do computerized systems automatically lockout a user after three failed attempts at logon?

Table 9.5 (Continued)

6		**Review of IT Infrastructure and Software**		
	6.1	Server Rooms		
			The room is secure	
			IT access only	
			Tidy and in good working order.	
			Has back-up and disaster recovery procedures in place	
			Date/time functionality of servers are correct	
			If data files are stored on a server, they cannot be deleted or modified	
	6.2	Stand-Alone Instruments Connected to PCs		
			Access to C: prompt restricted	
			Hard drive access and data files:	
				Determine if files are stored on the local c: drive
				If data files are stored on the local drive, they cannot be deleted or modified
			Date/time functionality is correct	
			Date/time cannot be changed by the lab personnel-IT Administrator only	
			Procedural controls exist for any hand-held controllers (e.g. Agilent Gameboys)	
			Flash drives and DVD drives are disabled	
			Files cannot be downloaded from the Internet	
			Antivirus software installed, current and operational	
	6.3	Client-Server Instruments Connected to PCs		
			Data cannot be copied to and from folder on server	
			Data cannot be placed in a different folder to "hide" it (e.g. "trial injections)	
			Time locked on server system and cannot be accessed or changed	
			Firewall and Antivirus and software installed, current and operational	
			Chromatographic Data Systems (CDS) files cannot be altered	
			Remote instrument controls are restricted or controlled	
	6.4	Computer Software Purchase, Installation, Validation and Maintenance		
			Purchase, installation and validation performed using URS and computer validation protocol	
			Installation of version upgrades, security upgrades and patched, service packs, etc. are current	

(Continued)

Table 9.5 (Continued)

7			Storage of Data and Notebooks
	7.1	☐	Are returned notebooks and documentation properly recorded, stored, and archived?
	7.2	☐	Are returned notebooks indexed?
	7.3	☐	Is there a system in place which tracks notebook index topics?
	7.4	☐	Is there a validated system in place to optical scan and archive laboratory notebooks and records?
	7.5	☐	Are copies of archived electronic data stored off-site?
	7.6	☐	Are both on-site and off-site data protected from fire, water, and other environmental hazards?
	7.7	☐	Is data and record destruction carried out per Corporate policy and Site SOPs?

The example notebook and data review checklist is provided in electronic format in the Chapter 9 Appendix (www.wiley.com/go/Bliesner/LabControl_GMPEnvironment).

  - **Confirm** the performance of work steps or tasks with physical verification of outputs or results within mapped processes (e.g. inspection of cleanliness of glassware, proper labeling of samples and solution glassware, etc.).
  - **Challenge** the operation of selected LCS sub-systems within mapped processes (e.g. perform data retrieval from an archive, work back from published and reported data to raw data including metadata, try to modify or erase electronic records, entry into control areas, etc.).
  - **Witness** execution of compliance related sub systems as circumstances arise (e.g. Performance of Deviation or Laboratory investigations in real-time as they occur, response to power failures, response to equipment failures such as stability chamber alarms, etc.).
- Creating and implementing Audit Plans and integrating the audits into the regular, routine QA oversight function (Figure 9.7).

Employee Awareness and Training

It is of paramount importance that all personnel within a pharmaceutical company be aware, and be trained on the principals of data governance and data integrity. The most effective means to accomplish this is through direct management involvement in the preparation and presentation of training materials related to data governance and data integrity.

The following outline servers as a starting point for management design of a Data Governance and Data Integrity training program. Please note that employee involvement in the creating and presentation of portions of the training is also a key factor in success of the training.

1. The History of Data Integrity and Data Governance in the Pharmaceutical Industry
2. Regulatory Expectations Regarding Data Integrity
   a. USFDA
   b. MHRA
3. Definition of Data Governance and Data Integrity
4. The Importance of Data Integrity
5. Your Company Policies and Procedures Regarding Data Governance and Data Integrity
   a. Individual Review of Data Governance Policy or Procedure
   b. Individual Review of Data Integrity Policy or Procedure
   c. Group Discussions Regarding SOPs Which Provide the Backbone of the Data Governance and Data Integrity System
   d. Knowledge Checks and Discussions

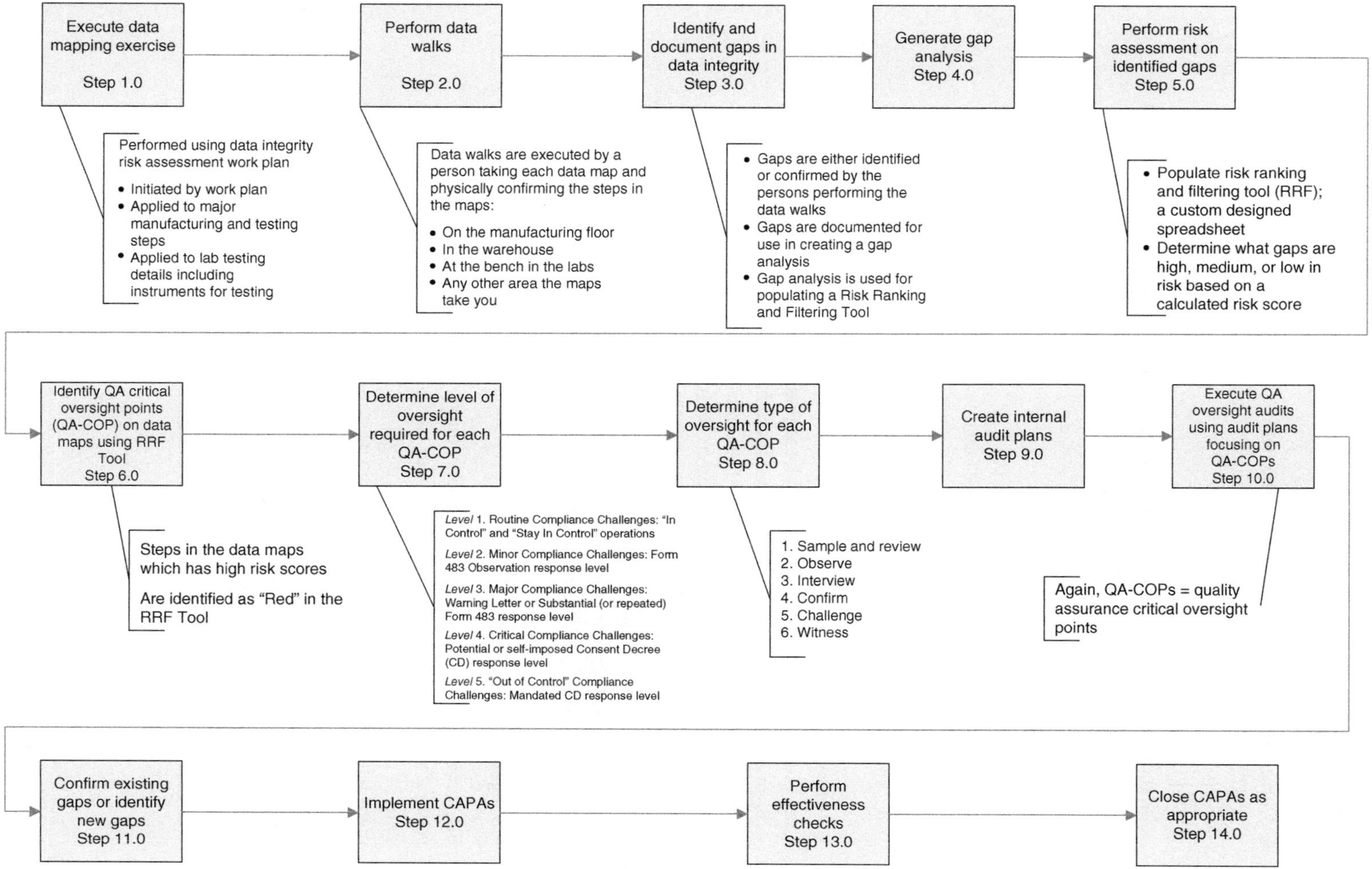

Figure 9.7 Quality assurance data and operations audit program work flow.

6. Current events: Observations from the field
   a. USFDA Warning Letters Citing Issues Related Data Integrity
      i. Researched by Employees
      ii. Presented by Employees
   b. Industry News Related to Data Integrity (Researched and Presented by Employees)
      i. New Services (e.g. FiercePharma family of news services, www.fiercepharma.com/)
      ii. Industry Publications
      iii. FDA Warning Letter e-mail Service
   c. Data Integrity Challenges with the Organization (as appropriate)
7. Employee Data Integrity Confidential Reporting Mechanisms
   a. Data Integrity "Hotline" or Similar Reporting System
   b. Amnesty Program for Past Transgressions
8. Timing and Timelines Regarding Data Governance and Data Integrity Instruction
   a. Full Course Instruction Upon Implementation of the Training Course
   b. Full Course Annually per GMP Requirements
   c. Quarterly Updates and Employee Presentations
   d. On-Demand Instruction as Conditions Warrant

Details of some possible content have been previously discussed within the body of this chapter and the text at large. However, each organization will have their own unique requirements for content and are encouraged to modify or enhance this outline to suite their individual needs.

Management Oversight

FDA is very clear in its position that Management's involvement is essential in preventing and correcting conditions that can lead to data integrity problems. They state in the most recent data integrity guidance that "It is the role of management with executive responsibility to create a quality culture where employees understand that data integrity is an organizational core value and employees are encouraged to identify and promptly report data integrity issues. In the absence of management support of a quality culture, quality systems can break down and lead to CGMP noncompliance" [1]. Although it is imperative that executive management sets the tone, management at all levels (e.g. Site, Department, Group, Section, and Bench-Level) must create and implement a quality culture to prevent lapses in data integrity.

Some of the steps or actions that management can take to enhance their chances of success in the implementation of the quality culture are to:

- Develop a functional understanding of the Production System, including understanding where risks to data integrity are most likely to occur (see Section "Risk Identification, Ranking, and Filtering").
- Similarly, develop a functional understanding of the LCS, including understanding where risks to data integrity are most likely to occur (see Section "Risk Identification, Ranking, and Filtering"). Historically non-laboratory management, including Site Leads, do not understand or appreciate the complexities of the LCS.
- Be aware of the balance (or lack of balance) between workloads, the level of qualified and experienced staffing.
- Be aware of the balance (or lack of balance) between workloads and the appropriateness of Production and Laboratory Facilities and Equipment.
- Use the output from the Data Mapping and Data Walking sessions to identify, rank and prioritize, address, and mitigate the risks to Data Integrity on the manufacturing floor and at the laboratory bench, through corrective and preventive action plans.
- Stay current with the most recent regulatory actions regarding data integrity issues through sources like the FDA's weekly Warning Letter free e-mail subscription service, industry blogs, industry trade magazines, etc.
- Meet regularly with employees to train and discuss current trends related to data integrity violations, FDA expectations, changes in corporate expectations or site policies and procedures related to data governance and data integrity.
- Periodically visit the manufacturing floor and the laboratory bench to keep connected with employees and observe work to convey your interest in, not only data integrity, but your value of the work they undertake.
- Avoid counterproductive behavior such as:
  - Putting pressure on personnel to release product to market, even if they feel it is not up to quality standards.
  - Displaying a lack of appreciation for the complexity and challenges of LCS operations.
  - Making employees feel that they may be terminated if they ever challenge their leaders; instead, encourage open and honest discussions.
  - Making employees feel that mistakes will be held against them, and will jeopardize any chance of advancement.
  - Making employees feel they will not be treated with respect if they make honest mistakes.

- Appreciate and understand cultural differences and norms in ethnically diverse work environments, and anticipate how they may impact data governance and data integrity.
- Establish a data integrity "hotline" or similar system reporting system which will allow employees to anonymously report data integrity issues or concerns.
- Ensure that the right number of the right people, have the right tools, and are working the right work environment which values their personal safety.
- Stay engaged with QA so they understand that you understand the value and need for a robust quality culture.

Bottom line is, if you lead your people well with honesty and integrity, they will follow you wherever you want to take them and do it in a productive, compliant and satisfying manner. The choice is yours.

Tools and Templates

The following are provided in electronic format in Example Tools and Templates, which can be found in the Chapter 9 Appendix (www.wiley.com/go/Bliesner/LabControl_GMPEnvironment):

- Example-Template Data Integrity Risk Assessment Work Plan
- Example-Template Data Map for the Manufacture of a Drug Product
- Example-Template A Generic Laboratory Instrument Data Map
- Example-Template Gap Analysis for Manufacturing Steps
- Example-Template Gap Analysis for a Laboratory Instrument Data Map
- Example-Template Risk Ranking and Filter Tool
- Example-Template Laboratory Data Flow for Notebook and Data Review
- Example-Template Notebook and Data Review Checklist
- Example-Template Quality Assurance Data and Operations Audit Program Work Flow

Glossary

Audit Trail A secure, computer-generated, time-stamped electronic record that allows for reconstruction of the course of events relating to the creation, modification, or deletion of an electronic record. The audit trails capture the "who, what, when, and why" of a record.

Backup Files A true copy of the original record that is maintained securely throughout the record retention period. Backup data must be complete, exact, and secure from alteration, inadvertent erasures, or loss. The backup file should contain the data (which includes associated metadata) and should be in the original format or in a format compatible with the original format. FDA's use of the term backup is consistent with the term archive.

Data Facts, figures and statistics collected together for reference or analysis. All original records and true copies of original records including, source data and metadata and all subsequent transformations and reports of these data, which are generated or recorded at the time of the GMP activity or process and allow full and complete reconstruction and evaluation of the GMP activity. Data may be contained in paper records (such as worksheets, forms and logbooks), electronic records and audit trails, photographs, audio or video files or any other media whereby information related to GMP activities is recorded.

Data Criticality Data has varying importance to quality, safety, and efficacy decisions. Therefore, data criticality may be determined by considering how the data is used to influence the decisions made.

Data Governance The sum total of arrangements to ensure that data, irrespective of the format in which it is generated, is recorded, processed, retained, and used to ensure a complete, consistent, and accurate record throughout the data life cycle.

Data Integrity The extent to which all data are complete, consistent, and accurate throughout the data lifecycle.

Data Lifecycle All phases in the life of the data (including raw data) from the initial generation and recording through processing (including analysis, transformation, or migration), use, data retention, archive, retrieval, and destruction. Facts, figures, and statistics collected together for reference or analysis. All original records and true copies of original records including source data and metadata and all subsequent transformations and reports of these data, which are generated or recorded at the time of the GMP activity or process and allow full and complete reconstruction and evaluation of the GMP activity. Data may be contained in paper records (such as worksheets, forms, and logbooks), electronic records and audit trails, photographs, audio or video files, or any other media whereby information related to GMP activities is recorded.

Data Processing A sequence of operations performed on data to extract, present, or obtain information in a defined format. An example includes transforming a raw electronic signal from an HPLC detector into a chromatogram, which is then used to generate a numerical result.

Data Risk Data risk represents the vulnerability to unauthorized deletion or alteration of data and the likelihood of detection during routine review.

Dynamic Record A record format that allows interaction between the users and the record content. For example, a dynamic chromatographic record may allow the user to change the baseline and reprocess chromatographic data so that resulting peaks appear smaller or larger. It may also allow users to modify formulas or other entries used to calculate test results.

Electronic Signature A signature in digital format that represents the person signing the document or record. Electronic signatures are legal equivalents of handwritten signatures.

Metadata Metadata is the contextual information required to understand data and describes the attributes of other data and provides context and meaning. Metadata is data about data. For example, the number "9" is meaningless without metadata, such as an indication of the unit "mg." In the example 9 mg, the mg is the metadata describing the data "9."

Raw Data Original records retained in the format in which they were originally generated (for example, paper or electronic) or as a "true copy." Raw data must be contemporaneously and accurately recorded by permanent means.

Static Record A fixed-data document such as a paper record or an electronic image. A static record allows little or no interaction between the user and the record content. An example of a static record can be the printout from a balance, where once printed there is no ability to electronically interact with the printout.

True Copy A copy of original information verified as an exact (accurate and complete) copy having all of the same attributes and information as the original. The copy may be verified by dated signature or by a validated electronic signature. A true copy may be retained in a different electronic file format to the original record, if required, but must retain the equivalent static/dynamic nature of the original record.

References

1. USFDA Draft Guidance to Industry titled (2018). Data Integrity and Compliance With CGMP, December 2018.
2. Medicines & Healthcare products Regulatory Agency (MHRA) (2018). 'GXP' Data Integrity Guidance and Definitions, March 2018.
3. Pharmaceutical Inspection Convention Pharmaceutical Inspection Co-Operation Scheme (PIC/S) Guidance titled (2018). Good Practices for Data Management and Integrity in Regulated GMP/GDP Environments, November 2018.

4 Garry Wright (2016). Preparing for a Data Integrity (DI) Audit. European Laboratory Compliance Specialist, Agilent Technologies, 2 February.

Further Reading

McDowall, R.D. (2019). *Data Integrity and Data Governance, Practical Implementation in Regulated Laboratories*. R.D. McDowall, Ltd., Published by the Royal Society of Chemistry www.rsc.org.

Parenteral Drug Association (PDA) (2018). Data Integrity Management System for Pharmaceutical Laboratories. *Technical Report No. 80*. MD, USA.

World Health Organization (2003). Application of Hazard Analysis and Critical Control Point (HACCP) methodology to pharmaceuticals, Annex 7. *WHO Technical Report Series No. 908*.

10

Components of the Stability Program Sub Element (SB)

Description of the Stability Program Sub Element

The purpose of stability testing is to provide evidence of how the quality of an Active Pharmaceutical Ingredient (API) or drug product varies with time under the influence of a variety of environmental factors such as temperature, humidity, and light. This information is obtained by implementing a clearly defined and well-established stability program. The stability program also includes the study of product-related factors that influence product quality, for example, interaction of API with excipients, differences in container closure systems and packaging materials. As a result of stability testing, a re-test period for the API or a shelf-life for the drug product can be established and storage conditions can be recommended [1–3].

It should also be mentioned that the establishment and maintenance of a stability program is explicitly required in the current good manufacturing practice regulations. Section 211.166 of the GMPs states "(a) There shall be a written testing program designed to assess the stability characteristics of drug products. The results of such stability testing shall be used in determining appropriate storage conditions and expiration dates." [1]

Depending upon the size of the organization and the number and diversity of product types, establishing and maintaining the stability program may be a significant undertaking. The stability program may require its own department within the company, employing a substantial staff, occupying a large facilities footprint, and include significant numbers and types of instrumentation and equipment. For smaller organizations, the stability program may be reduced in size and complexity and may even be partly or completely outsourced. Regardless, there are certain regulatory expectations with respect to the design, and operation of a stability program.

In general, the Stability Program sub element usually includes at least 10 individual topics. These topics are as follows: (i) stability program overview

Laboratory Control System Operations in a GMP Environment, First Edition. David M. Bliesner.

Companion website: www.wiley.com/go/Bliesner/LabControl_GMPEnvironment

and work flow description, (ii) stability study initiation (requests and protocol generation), (iii) stability study activation (sample set-down), (iv) time zero testing, review, and reporting, (v) sample pulls, (vi) sample testing and reporting of results, (vii) data analysis, reporting, and shelf-life determination, (viii) stability program administration, (ix) additional types of stability studies, and (x) stability chambers.

Contents of the Sub Element

A list of example standard operating procedures, which define, establish, and maintain a model stability program, is outlined in the succeeding text [1–7]. It should be noted that the outline represents a model stability program. Therefore, the outline and content of each SOP will need to be tailored to the individual organization's requirements based on product mix, production volume, and overall organizational structure and function.

Model Standard Operating Procedures for Establishing and Maintaining a Stability Program

1. Procedure #1: An Overview of the Stability Program
   - 1.1 Purpose of the stability program
   - 1.2 Roles and responsibilities
   - 1.3 Job descriptions
   - 1.4 Work flow
2. Procedure #2: Initiating Stability Studies (Study Requests and Protocol Generation)
   - 2.1 Reasons for initiating stability studies
   - 2.2 Protocol design and creation
   - 2.3 Protocol review and approval
   - 2.4 Protocol implementation
   - 2.5 Protocol modification and change control
   - 2.6 Bracketing and matrixing considerations
3. Procedure #3: Activating Stability Studies (Sample Set-Down)
   - 3.1 Sampling from manufacturing line and temporary sample storage
   - 3.2 Sample transport from manufacturing to temporary storage area
   - 3.3 Sample receipt, inspection and inventory
   - 3.4 Stability sample "for study" labeling
   - 3.5 In-chamber sample placement
   - 3.6 Stability pull and testing schedule
   - 3.7 Sample tracking and inventory control

4. Procedure #4: Time Zero Testing, Review and Reporting
   4.1 Definition of time zero testing
   4.2 Developmental or investigational studies
   4.3 Annual stability studies
   4.4 Data reporting mechanisms for study inclusion
   4.5 Limitations on use of release data for time zero
5. Procedure #5: Withdrawal of Samples from Stability Chambers (Sample Pull)
   5.1 Sample pull date notification
   5.2 Quantities and conditions
   5.3 Removal from chambers with prescribed windows and documentation of removal
   5.4 Temporary sample storage (pre-transport/pre-test)
   5.5 Notification to laboratory of testing
   5.6 Transfer of samples to testing laboratory
   5.7 Storage of samples at testing laboratory
   5.8 Deviations for missed pulls or time-windows
   5.9 Periodic physical inventory of samples in chambers
6. Procedure #6: Sample Testing and Results Reporting
   6.1 Types and appropriateness of performed tests
   6.2 Additional tests to establish and/or evaluate drug product stability
   6.3 Use of appropriate testing specifications
   6.4 Use of validated or verified test methods
   6.5 Use of certified reference standards and reference materials
   6.6 Use of qualified (IQ/OQ/PQ) and calibrated equipment
   6.7 Use of validated software systems
   6.8 Analysts training and qualifications
   6.9 Testing and QA review turnaround requirements
   6.10 Data capture, review, and handling procedures
   6.11 Deviations for missed review time-windows
   6.12 Handling of Out-of-Specification results
   6.13 Addressing Out-of-Trend results
7. Procedure #7: Stability Data Analysis, Reporting, and Shelf-Life Assignment
   7.1 Work flow for compiling, analyzing, and reporting stability data
   7.2 Periodic analysis and reporting of results
   7.3 Early termination or cancelation of stability studies
   7.4 Identifying, tracking and reporting events or trends
   7.5 Shelf-Life estimation and assignment
   7.6 Input to stability data into the Annual Product Review (APR)
8. Procedure #8: Administration of the Stability Program and Stability Studies
   8.1 Keeping current with FDA and ICH expectations for stability studies and expiration date assignment
   8.2 Review and update of stability related SOPs

   8.3 Monitoring and maintaining testing schedule, sample-pulls, testing, and reporting
   8.4 Study and protocol amendment management
   8.5 Post application approval change management
   8.6 Third party (outsourced) study management
   8.7 Study protocol deviation management
   8.8 Stability related investigation procedures (OOS and OOT)
   8.9 Study termination procedures
   8.10 Study completion and closure procedures
   8.11 Annual product stability set-down (annual addition)
   8.12 Sample destruction and reconciliation
   8.13 Training and qualification of stability program management personnel
9. Procedure #9: Additional Types of Stability Studies
   9.1 Knowledge of Drug Substance and Drug Product to support stability studies
   9.2 Shipping studies
   9.3 Thermal excursion studies (high and low temperatures)
   9.4 Bulk hold time studies
   9.5 Intermediate or blend hold time studies
   9.6 Post approval change studies
   9.7 Collection and proper storage of retain (reserve) samples for each batch
10. Procedure #10: Selection, Qualification, Calibration, and Maintenance of Stability Chambers
   10.1 Stability chamber user requirements specifications
   10.2 Stability chamber qualification
   10.3 Chamber security and access control
   10.4 Alarm and monitoring systems and practices
   10.5 Chamber excursions, back-up, and disaster recovery
   10.6 Periodic calibration and preventive maintenance
   10.7 Periodic chamber sample inventories and reconciliation

The first nine topics listed in the introduction of this chapter and outlined in the model SOP list in the previous text are described in additional detail, along with some notes and clarifications as appropriate in Table 10.1, "Components of the stability program." The list of items under each topic is fairly comprehensive but may not be totally inclusive of all those components, which may constitute the sub element. Individual organizational requirements will vary.

A separate section for topic 10, which specifically addresses stability chambers, is included and described in Table 10.2 "Selection, qualification, calibration, and maintenance of stability chambers."

Table 10.1 Components of the stability program.

#	Stability program sub element topic	Notes and clarifications
10.1	**PROCEDURE #1:** An Overview of the Stability Program	
10.1.1	***Is the Purpose of the stability program:*** • To determine the re-test requirements, shelf-life, and subsequent expiration dating of API and drug product? • To monitor the stability (e.g. quality) of commercial product? • To address regulatory inquires, requests, or commitments? • To address and support manufacturing deviations? • To support changes in drug product formulation or container closure systems? • To support changes in the manufacturing process or location of manufacturing? • To determine API or drug product stability for different climatic zones than included in the original API or drug product regulatory filing? • Evaluation of stability for reworked API or product? • To determine appropriateness of shipment of samples according to defined transport condition? • To determine bulk hold-times for finished product? • To determine in-process hold time as part of the manufacturing process controls? • To determine stability of clinical materials to support clinical trials?	

(Continued)

Table 10.1 (Continued)

#	Stability program sub element topic	Notes and clarifications
10.1.2	***Are the Roles and Responsibilities in regards to the stability program, of the following parties defined in the overview procedure:***	
	• Research and development including:	
	o Formulation development and product development?	
	o Analytical research and development?	Includes method development and validation.
	o Process development and process validation or verification?	
	o Contract research and development?	
	o Client R&D?	This pertains to organizations which provide contract stability services which need to interface with their client's research and development department.
	• Quality control laboratory?	Both internal and external. For example, some third-party stability services involve only storage of stability samples and does not include QC testing, which is done in-house.
	• Technical services group?	
	• Stability program management group?	
	• Quality assurance group?	
	• Regulatory affairs group?	
	• Contractors including:	
	o Routine analytical testing services?	
	o On-going stability services?	This can include stability sample storage at various climatic conditions, testing of stability samples or both.
	o Manufacturing services?	
	o Packaging services?	
	o New formulation and product development services?	
	• Company sites or groups within the company network?	This applies to large organizations with multiple sites or groups. These sites may be US based or offshore sites.

Table 10.1 (Continued)

#	Stability program sub element topic	Notes and clarifications
10.1.3	***Do job descriptions exist for the in-house positions which are described above?***	In the past, job descriptions were primarily considered as human resource related requirements. However, formal job descriptions are increasingly being viewed by regulators as a GMP requirement. In addition, US FDA has cited organizations for inadequate staffing resources to support stability programs. Therefore, companies' need to maintain an adequate staffing level, of qualified personnel, to ensure proper oversight and execution of API and drug product stability programs.
10.1.4	***Do work flow diagrams, flow-charts, or process diagrams exist which describe how work is executed during the operation of the Stability Program?*** ***If yes, do they:*** • Describe the linkage to responsible parties? • Link to SOPs which provide guidance and detail on execution of each step?	
10.1.5	***If work-flow diagrams, flow-charts, or process diagrams do not exist, are there procedural controls in place such as:*** • Tabularized list or matrix of work steps which show the following: o Required tasks for each step number? o A listing of persons responsible for each step? o Definition of what is required for each step? • A listing or matrix of procedures (SOPs) which provide guidance for each step in the stability program work flow?	

(Continued)

Table 10.1 (Continued)

#	Stability program sub element topic	Notes and clarifications
10.1.6	***Does the Stability Program include, at a minimum, the following specific work flow steps:***	
	• Requests for stability studies?	That is, a formal, written, request to initiate a stability study.
	• Assessment of what is needed in the study versus a standard study protocol template?	A stability study protocol template is the backbone for the design of a stability study. The structure and content of stability study protocols and reports is discussed in Procedure #2: Initiating Stability Studies: Study Requests and Protocol Generation.
	• Drafting the stability protocol?	
	• Review of the protocol by all responsible parties?	
	• Manufacturing, sampling, packaging, and delivery of samples for set-down?	Set-down is a term used that means the placement of stability samples in stability chambers.
	• Temporary sample storage?	A means to document and ensure sample chain of custody should be implemented and include a temporary storage location for stability samples following delivery of samples following sampling from the manufacturing line.
	• Sample set-down in chambers?	
	• Sample pull from chambers per testing schedule?	Removal of stability samples from the stability chamber at the time point defined in the stability protocol is referred to as a sample pull.
	• Chamber inventory update?	An accurate accounting of the inventory of stability samples that are contained in the stability chamber must be current and carefully maintained. This inventory changes as samples are added or removed from the chamber.

Table 10.1 (Continued)

#	Stability program sub element topic	Notes and clarifications
	• Sample transfer from the temporary storage location to storage within the QC laboratory?	Samples should be stored in a secure, limited access location within, or nearby, the QC laboratory prior to and during testing. Samples are removed from this location and destroyed only after the test results have been generated, reviewed, entered into the stability data reporting system and all investigations or deviations, which may have risen have been completed.
	• Testing of samples and documentation of results?	Performed by the QC laboratory or outside contract testing laboratory. Documentation of the testing performed and results are normally contained in laboratory notebooks, in worksheets, and in the electronic records.
	• Initial review and evaluation of data generated during testing?	Performed by the QC laboratory. This involves comparing results obtained during testing to predetermine product specifications.
	• As necessary, perform investigations (e.g. OOS or OOT) and take appropriate action?	OOS or Out-of-Specification and OOT or Out-of-Trend investigations must be initiated when QC testing of stability studies produce an out of specification result or when the result is not in line with the previous stability point (e.g. dramatic change in appearance, a change in assay value, which is ±5% from the initial assay value). Appropriate action may include issuance of a Field Alert Report (FAR), Study Termination, Recall, etc. as necessary.
	• Reporting the results?	This normally involves populating a standardized stability report template with the results obtained during testing.
	• Performing data analysis?	This is to determine acceptability of results and determine if trends are developing over time with respect to the quality attributes of the API or drug product.

(Continued)

Table 10.1 (Continued)

#	Stability program sub element topic	Notes and clarifications
	• Completing the stability report?	
	• Deriving conclusions from the results?	These conclusions may include items such as: • Selection of appropriate storage conditions and instructions for API and marketed products. • Assignment of shelf-life. • Assignment of a rest period for API. • Selection of packaging material. • Determination of appropriate transport conditions. • Bulk hold times, etc.
	• Finalize, review, and approval of the stability report? • Publishing and distributing the stability report?	
10.2	**PROCEDURE #2:** Initiating Stability Studies: Study Requests and Protocol Generation	
10.2.1	***Does the procedure define or describe the reasons or circumstances for initiating stability studies, including:***	
	• Generating data for submissions (NDA or ANDA registration batches)?	
	• Annual product stability set-down?	At least one lot of each manufactured drug product must be placed on stability annually.
	• Developmental (R&D) work (internal requests)? • Developmental (R&D) work (external requests)? • In response to product quality changes observed in previous stability studies (e.g. uptake in moisture, color changes, changes in dissolution profiles, etc.)?	
	• Post approval stability commitment to regulatory agencies?	Sometimes referred to as SUPAC-scale-up and post approval changes.
	• First three commercial batches (also serve as process validation batches for ANDAs)?	

Table 10.1 (Continued)

#	Stability program sub element topic	Notes and clarifications
	• Additional, specific regulatory requests?	
	• To support compliance investigations (e.g. Due to deviations in manufacturing process during commercial production)?	There may be instances where deviations from the manufacturing process may raise questions with respect to the long-term stability of finished product or API. As a precautionary measure, samples from the impacted lots may be placed on stability so that impact of the deviation can be monitored over time.
	• Bulk hold-time studies?	
	• Thermal excursion studies (e.g. In support of product development or in response to actual excursions)?	
	• Change in manufacturing site?	
	• Change in API manufacturing process at current supplier or a change in the actual API supplier?	
	• Change in drug product container-closure system?	
	• Change in drug product formulation?	
	• Change in drug product manufacturing process?	
	• To support reworked or reprocessed batches?	
	• Exceeding yearly volume expectations?	As mentioned earlier, at least one batch per year should be placed on stability. However, if production exceeds typical manufacturing volume, additional lots of product should be place on stability. This requirement should be pre-determined and codified in an SOP.
	• To support shipping studies?	
	• "Other" reasons for initiating stability studies?	Any event or change that may impact product quality over time, should initiate the discussion regarding whether information gained by an additional stability study may be of value.

(Continued)

Table 10.1 (Continued)

#	Stability program sub element topic	Notes and clarifications
10.2.2	***Does the procedure describe the design, creation and modification of the following:***	
	• A master stability protocol template?	See Chapter 10 Appendix (www.wiley.com/go/Bliesner/LabControl_GMPEnvironment) for an example stability study protocol template.
	• Study specific protocols?	Such as stability studies to support manufacturing deviation, addition of a container closure system, etc.
	• Protocol "minimum information to contain" checklist, which includes:	These are the main components of a typical stability protocol.
	o Organization initiating the protocol?	
	o Product name?	
	o Strength and dosage form?	
	o Protocol number or identifier?	
	o Titles, signatures and review and approval dates including management (e.g. Stability, Project Management, Regulatory, Customer (e.g. requester), and Quality Assurance (QA))?	
	o Purpose or objective of the study?	
	o Batch number?	
	o Batch size?	
	o Expiration date?	
	o Manufacturer?	
	o Manufacturing site?	
	o Manufacturing date?	
	o Raw material (API) batch number?	
	o Raw material (API) manufacturing site?	
	o Packager?	
	o Packaging site?	
	o Packaging date?	
	o Packaging components (e.g. description of primary packaging)?	

Table 10.1 (Continued)

#	Stability program sub element topic	Notes and clarifications
	o Sampling plan or policy? o Sample management? o Container-closure orientation (e.g. upright, inverted, or horizontal)? o ICH storage conditions and testing internals? o Analytical procedures? o Testing specifications? o Testing data management procedure references? o Data analysis and reporting procedure references? o Out-of-Trend (OOT), Deviation, Out-of-Specification (OOS) and Field Alert procedure references? o Protocol amendment and discontinuation procedure references? o Stability study start date (e.g. in-chamber placement)?	
	• Compliance with ICH Guidelines, including:	ICH is the International Council for Harmonization, which was formerly known as the International Conference on Harmonization.
	o Temperature and humidity conditions for desired market? o Testing intervals? o Inclusion of Intermediate (30 °C/65% RH) and Accelerated (40 °C/75% RH) in addition to long-term (25 °C/60% RH) conditions for the US market?	Other conditions may be appropriate depending on the product and the container-closure system. See a copy of the ICH harmonized tripartite guideline titled "STABILITY TESTING OF NEW DRUG SUBSTANCES AND PRODUCTS Q1A (R2) in the Chapter 10 Appendix (www.wiley.com/go/Bliesner/LabControl_GMPEnvironment)

(Continued)

Table 10.1 (Continued)

#	Stability program sub element topic	Notes and clarifications
10.2.3	***Does the procedure describe the process of reviewing and approving stability protocols?***	
10.2.4	***Does the procedure describe how stability protocols are formally implemented?***	
10.2.5	***Does the procedure describe how stability protocols can be modified with modifications being formally captured under a change control system?***	
10.2.6	***Does the procedure include a discussion of bracketing or matrixing?***	Bracketing is the design of a stability schedule such that only samples on the extremes of certain design factors (e.g. strength, package size), are tested at all time points as in a full design. See ICH Q1A (R2). Matrixing is the design of a stability schedule such that a selected subset of the total number of possible samples for all factor combinations is tested at a specified time point. See ICH Q1A (R2).
10.3	**PROCEDURE #3:** Activating Stability Studies (Sample Set-Down)	
10.3.1	***Is there a procedure in place which provides instructions and guidance with respect to the following for activating stability studies:*** • Sampling product from the manufacturing line including the following considerations? o A representative sampling from the manufactured batch? o Sampling throughout the manufacturing run? o Sampling of proper and sufficient number of samples to support testing, re-testing, and investigations at each condition and station?	

Table 10.1 (Continued)

#	Stability program sub element topic	Notes and clarifications
	• Sampling is documented on the batch record including time, date location, sampling throughout the manufacturing run, name of the individual performing sampling, and confirmation of random/representative selection of samples? • Documentation supporting a sampling chain of custody for transfer to stability program temporary storage location?	Also make sure that no "cherry picking," that is purposeful selection of samples from the line.
10.3.2	***Is there a procedure in place which provides instructions and guidance with respect to the following for transport of samples from manufacturing to the Stability Program temporary storage area, which includes:*** • Description of "within facility" transport and storage including proper chain of custody, designated storage locations, designated temporary storage locations, and assurance that samples are stored in accordance with drug product labeling, prior to set-down on stability? • Description of "between facilities" transport and storage to include proper chain of custody, temperature, and humidity monitoring during transport, a shipping study to support shipping conditions and designated temporary storage locations, and assurance that samples are stored in accordance with drug product labeling prior to set-down on stability?	

(Continued)

Table 10.1 (Continued)

#	Stability program sub element topic	Notes and clarifications
10.3.3	***Are procedures in place which, describe sample receipt, inspection and inventory at the temporary storage location prior to set-down on stability, and do they include the following:***	
	• Final marketed presentation including secondary packaging?	
	• Simulated bulk container if shipped in bulk?	
10.3.4	***Are procedures in place that describe labeling of stability samples "for study," which include:***	
	• Full explanation of all study information required on the label including product description, lot, storage condition, orientation, date placed in chamber, and date to be removed from chamber?	
	• Requirements to label the primary packaging (container-closure) system?	
	• Requirements to label the secondary packaging?	
	• Requirements to label additional storage containers like shipping containers or storage bins?	In chamber storage can include storage bins, trays, etc. versus simply place samples on shelving.
10.3.5	***Are procedures in place that describe details of in-chamber sample placement including:***	
	• Description of the stability study tracking and reporting system which may include or require:	
	o A data system requiring manual entries?	Such as recording in a logbook or log sheet.
	o A data system using automated entries?	Such as a barcode scanning system.
	o A hybrid data entry system?	

Table 10.1 (Continued)

#	Stability program sub element topic	Notes and clarifications
	• Instructions for creating or inputting the following items into the stability study tracking system: o Record creation and data entry? o Sample location? o Quantity? o Conditions? • Final confirmation that set-down conditions are consistent with labeled storage conditions and expected product shipping/ distribution conditions?	
10.3.6	***Are procedures in place that describe details of the stability pull and testing schedule to include:*** • Pull and testing calendar creation? • Pull and testing calendar modification? • Pull and testing calendar schedule monitoring which may include: o A manual calendar requiring manual tracking and notifications? o An automated, electronic calendar system which tracks and notifies personnel when pulls and testing are required?	A stability pull is a scheduled removal of stability samples form stability chambers.
10.3.7	***Are procedures in place that describe details of how the location stability samples are physically tracked and how chamber inventories are maintained and confirmed throughout the study to include:*** • Continuous monitoring of chamber inventories and sample locations as testing progresses or a need to move samples from one chamber to another occurs?	

(Continued)

Table 10.1 (Continued)

#	Stability program sub element topic	Notes and clarifications
	• Periodic, independent chamber content inventory including:	Independent means the inventory should be conducted by persons not directly involved in stability sample handling and testing. This should preferably be performed by QA. Inventories should be conducted at least once a year.
	o A complete chamber inventory for small volume chambers?	Small volume means chambers that hold a limited number of samples versus room-sized chambers.
	o Statistical selection of samples for very large chambers?	It is in many cases impractical to perform full chamber inventories for room-sized chambers containing large numbers of samples.
10.4	**PROCEDURE #4:** Time Zero Testing, Review, and Reporting	
10.4.1	***Is there a procedure in place that defines time zero data for stability testing and how the time zero data are obtained and included in the stability study?***	Time zero data are the initial, starting point data for a stability study. Since product may be manufactured, packaged, held in bulk, etc. and not immediately placed on stability, the definition of time zero needs to be clearly defined in a procedure.
10.4.2	***Do development studies or investigational studies consider the following?*** • Developmental work may include generated data to support NDA, ANDA submission, formulation development, evaluating new container-closure systems, etc. • Investigation studies may include evaluating the impact of manufacturing deviations on drug product stability or deviations during drug product transport.	

Table 10.1 (Continued)

#	Stability program sub element topic	Notes and clarifications
10.4.3	***For annual stability studies, are the data generated at release of product to market and included on a Certificate of Analysis (CoA)?***	
	• Time zero data may come directly off the drug product release CoA	
	• Additional testing may be required even if the CoA data are used	For example, some products such as pediatric antibiotics may need to be reconstituted in water so they can be easily swallowed. This type of testing usually isn't performed at release.
10.4.4	***Does the procedure provide guidance on the mechanisms for inputting the data into the stability program for reporting which may include:***	
	• Data that may be entered manually, by transcription from the CoA into a database or paper system?	
	• Data that may be electronically transferred into a stability system database or other software used to analyze stability data?	
10.4.5	***Are there limitations of use of release data for time zero for inclusion into stability studies such as:***	
	• Mandating a maximum fixed time within which release data may be used as the time zero data for a stability study?	Standard industry practice usually mandates that full product retesting be performed if more than 30 days have passed since manufacture and if the product has not yet been placed on stability.
	• Mandating that drug product be tested independent of release data to generate time zero data for inclusion into a stability study?	

(Continued)

Table 10.1 (Continued)

#	Stability program sub element topic	Notes and clarifications
10.5	**PROCEDURE #5:** Withdrawal of Drug Product from Stability Chambers (Sample Pull)	
10.5.1	***Does the procedure address how sample pull date notifications are promulgated?***	This may include face-to-face notification after calendar review, e-mail notification, or automated software notification via e-mail.
10.5.2	***Does the procedure describe how to determine the quantities of samples to be pulled at each station to support all necessary testing and extra samples to support retesting and/or investigations as necessary?***	Station is another name for the sample storage condition (e.g. temperature and humidity) associated with a stability time-point.
10.5.3	***Does the procedure address removal from chambers within prescribed windows, and include the following specific points:***	Ensure that removal of samples from chambers is fully documented including chamber inventory updates (e.g. manual updates, bar code scan for electronic updates of the stability program management software, etc.) to ensure timely and proper testing and to maintain chamber inventories.
	• "Plus" windows only?	Never remove samples early from the chambers unless substantiated with scientific justification and practical reasons (e.g. site shut-downs, holiday schedules, etc.)
	• +3 to +5 days for 0 to 12-month samples stored at 25 °C/60% RH? • +1 week for 24-month (or longer) samples stored at 25 °C/60% RH? • Pull on-schedule for samples stored at accelerated or intermediate conditions (e.g. 30 °C/65% RH, 40 °C/75% RH)?	
10.5.4	***Does the procedure describe how samples are to be temporarily stored following removal from chambers and prior to transferring to the testing laboratory?***	This temporary storage should be secure with chain of custody tracking. Storage should also be at the recommended labeling conditions, such as control room temperature, refrigerated, etc.
10.5.5	***Does the procedure describe how the QC laboratory is notified that samples have been removed from the chambers and are ready for testing?***	This can take place manually (face-to-face) or by automated e-mail messaging via the stability management software.

Table 10.1 (Continued)

#	Stability program sub element topic	Notes and clarifications
10.5.6	***Is the process of transfer of samples to testing laboratory, including documentation of chain of custody, described in the procedure?***	
10.5.7	***Is storage of samples at the testing laboratory addressed in the procedure?***	Samples should be secured in a limited access location during testing.
10.5.8	***Does the procedure address deviations for missed pulls or time-windows?***	
10.5.9	***Does the procedure describe the requirement for periodic physical inventory of samples in chambers?***	Stability chamber inventory should be executed by protocol. In smaller chambers the inventory should be 100%. For larger chambers statistical sampling may be more appropriate. Inventory should be performed at least annually.
10.6	**PROCEDURE #6:** Sample Testing and Results Reporting	
10.6.1	***Does the procedure discuss the types and appropriateness of performed tests for stability samples (minimally) including:*** • Description? • Assay? • Impurities/Related Substances/Degradants?	
10.6.2	***Does the procedure also outline additional tests (as appropriate) to establish and/or evaluate drug product stability, including:*** • FTIR for ID?	
	• Moisture (by Karl Fisher or Loss on Drying)?	Sometimes Thermogravimetric Analysis (TGA) may be used.
	• Residual solvents?	Residual solvent measurement is typically restricted to API. However, if solvents that are identified as a risk are used during the manufacture of drug product, then testing should be performed for those solvents, at least on initial release of the product. See ICH Q3C(R6) Impurities: Guidelines for residual solvents.

(Continued)

Table 10.1 (Continued)

#	Stability program sub element topic	Notes and clarifications
	• pH?	
	• Weight variation?	For solid oral dosage forms (SODs).
	• Content uniformity?	
	• Friability?	
	• Hardness?	
	• Disintegration?	
	• UV/visible spectroscopy?	
	• Density or specific gravity?	
	• Melting point?	
	• Dissolution or drug release?	Drug release is dissolution testing for extended or controlled release drug products. That is drug products which are designed by formulation or physical structure to release the active ingredient of a period of time versus all at once (e.g. termed immediate release).
	• Microbial examination?	USP tests such as microbial enumeration and tests for specified organism and yeasts and molds, as appropriate.
	• Particulates?	Required for parenterals and other aseptically manufactured (e.g. sterile) products.
10.6.3	***Does the procedure discuss how testing specifications are generated, modified and updated and why the tests and specifications are appropriate?***	A testing specification delineates the types of tests which are required and a range of acceptable values which must be met for each test. Also, these specifications should make physical and chemical sense and provide data that supports the evaluation of product quality overtime.
10.6.4	***Does the procedure describe the required use of test methods which have been validated, or in the case of Compendial Methods, verified for their intended use via execution of validation or verification protocols with pre-approved acceptance criteria?***	These methods should also (as appropriate) be stability indicating. A stability indicating method has the ability to detect changes in drug product or drug substance quality over time.

Table 10.1 (Continued)

#	Stability program sub element topic	Notes and clarifications
10.6.5	***Does the procedure describe the required use of certified reference standards and reference materials?***	
10.6.6	***Does the procedure describe the use of qualified (IQ/OQ/PQ) and calibrated equipment?***	Equipment qualification should be performed using protocols with pre-approved acceptance criteria.
10.6.7	***Does the procedure require that software used to operate equipment, instruments or calculate results been properly validated including protocols with predetermined acceptance criteria?***	
10.6.8	***Does the procedure require that the analysts who perform testing to determine drug product or drug substance stability be properly trained and qualified or have been evaluated to possess the education, training, and experience or combination thereof to do so?***	Training and qualification are generally accepted as two levels of rigor with respect to analyst capabilities. Training may take the form of classroom instruction with subsequent knowledge checks. Qualification may take the form of demonstration by the analyst of their ability to perform more complex tasks (e.g. proof-of-proficiency).
10.6.9	***Does the procedure define the maximum amount of time allowed to complete testing of stability samples once they have been removed from stability chambers?***	Typically, 30 days is allowed for testing of samples stored under long-term storage conditions. For accelerated conditions, testing should be performed immediately after they are removed from the chambers.
10.6.10	***Does the procedure also define the maximum amount of time that is allowed for QA to review test results following testing of stability sample?***	Forty-five days is typically the maximum amount of time allowed for QA review.
10.6.11	***Does the procedure outline or flowchart the process of data capture, review, handling, and hand-offs as it relates to analysts, supervisors QA reviewers and the Stability Management Group for the analysis of stability samples?***	As discussed earlier, this may be captured in a separate procedure.

(Continued)

Table 10.1 (Continued)

#	Stability program sub element topic	Notes and clarifications
10.6.12	***Does the procedure require filing of a deviation with subsequent investigation in the event that the review times has been exceeded?***	
10.6.13	***Does the procedure reference an SOP for handling Out-of-Specification results?***	
10.6.14	***Does the procedure reference an SOP for addressing Out-of-Trend results?***	
10.7	**PROCEDURE #7:** Stability Testing Data Analysis, Reporting, and Shelf-Life Assignment	
10.7.1	***Does the procedure outline the work flow process for compiling, analyzing, and reporting stability data, and does it include the following steps:***	
	• Assignment of responsibilities and description of a system for the transfer of data from the QC laboratory for compilation and analysis by the Stability Management Group?	
	• If data transfer is manual, is there a means to verify transfer of the data into the stability program software/database?	Sometimes referred to as transcription accuracy verification (TAV).
	• If the data transfer is electronic, has the transfer of the data into the stability program software/database been validated?	
	• Reference to and periodic assessment of local procedure alignment with ICH Q1E "Evaluation of Stability Data"?	
10.7.2	***Does the procedure provide a requirement and guidance for periodically analyzing stability data and reporting throughout the entire stability study?***	Reporting should occur at each station (e.g. each station for time point) including the final station.

Table 10.1 (Continued)

#	Stability program sub element topic	Notes and clarifications
10.7.3	***Does the procedure provide a requirement and guidance for analyzing and reporting in cases of early termination or cancelation of a stability study?***	
10.7.4	***Does the procedure provide a requirement to identify, track and report unexpected trends observed throughout the study and include official notification to management of such unexpected events?***	
10.7.5	***Does the procedure provide guidance on how to establish initial shelf-life or extension of the original shelf-life including the following:***	Shelf-life is the time period during which a drug product or API is expected to remain within approved shelf-life expectations (e.g. maintains its quality attributes) provided it is stored under the conditions defined on the container label.
	• Presentation of data in an appropriate format?	Data are frequently presented in tabular, graphical, and narrative formats and/or combinations thereof.
	• Systematic evaluation of the data from stability studies, including assessment of any significant changes at the accelerated condition, intermediate conditions, and long-term stability conditions?	Significant change is minimally defined as a 5% change in assay value from the value obtained at the time zero testing. See ICH harmonized tripartite guideline titled "STABILITY TESTING OF NEW DRUG SUBSTANCES AND PRODUCTS" Q1A (R2) for additional discussion of "significant change."
	• Requirement to evaluate and report on each measured attribute?	This includes all the attributes from appearance, assay, hardness, etc.
	• Use of statistical analysis, as appropriate to establish re-test periods or shelf-life by analyzing quantitative attributes?	This involves the application of linear regression analysis to the quantitative results such as assay and degradation products. See ICH harmonized tripartite guideline titled "EVALUATION FOR STABILITY DATA" Q1E for a more detailed discussion.
	• Forgoing statistical analyses as deemed appropriate?	This may be possible when little or no change (or variability) is observed in the long-term stability study data and no significant change is observed at the accelerated storage condition.

(Continued)

Table 10.1 (Continued)

#	Stability program sub element topic	Notes and clarifications
10.7.6	***Does the procedure provide guidance on how to provide stability data for inclusion into the Annual Product Review (APR) including creating a stability summary report?***	
10.8	**PROCEDURE #8:** Administration of the Stability Program and Stability Studies	
10.8.1	***Does the procedure require periodic review of FDA and ICH guidance documents regarding stability studies and shelf-life determination?***	
10.8.2	***Does the procedure require periodic review and update (as appropriate) of stability program related SOPs?***	
10.8.3	***Does the procedure assign personnel to monitor and maintain sample-pulls, stability sample testing schedules, and reporting of results?***	
10.8.4	***Does the procedure assign personnel the responsibility for study and protocol amendment management?***	
10.8.5	***Does the procedure assign personnel the responsibility for post application approval change management including commitment to regulatory agencies and handling stability related scale-up and post approval changes?***	
10.8.6	***Does the procedure assign personnel the responsibility of management of third party (outsourced) stability studies?***	
10.8.7	***Does the procedure assign personnel the responsibility to identify and mange deviations from stability study protocols?***	

Table 10.1 (Continued)

#	Stability program sub element topic	Notes and clarifications
10.8.8	***Does the procedure assign personnel the responsibility for performing or assisting with the following stability related investigations:***	
	• Out-of-Trend identification and investigation support and follow-up?	
	• Stability sample testing Out-of-Specification investigation support and follow-up?	
	• Interface with the Field Alert Report (FAR) system for stability failures and ensure that FARs are filed with FDA within 3 days of discovery of the problem?	Specifically, 21 CFR 314.81 states "NDA – Field alert report. The applicant shall submit information of the following kinds [descriptions are listed later in the regulation] about distributed drug products and articles to the FDA district office that is responsible for the facility involved within 3 working days of receipt by the applicant."
	• Deviations from the stability program such as failure to withdraw samples within the prescribed time frame, failure to test stability samples within specified time frames, and shifting of samples from one chamber to another when equipment failures occur?	
10.8.9	***Does the procedure describe the necessary steps to terminate a stability study before the protocol is fully executed?***	
10.8.10	***Does the procedure describe the necessary steps to close a stability study upon its completion (e.g. stability protocol is fully executed)?***	

(Continued)

Table 10.1 (Continued)

#	Stability program sub element topic	Notes and clarifications
10.8.11	***Does the procedure describe the annual product stability set-down (annual addition) including the following considerations:***	
	• Placing at least one batch of each product per year on stability?	
	• Placing more batches on stability for higher volume products?	Products of which more than 50 lots per year have been manufactured are candidates for additional stability set-downs.
10.8.12	***Does the procedure describe steps for sample destruction and reconciliation at completion of testing at each station and end of the study?***	Reconciliation involves comparing what samples remain versus what should have been consumed in testing or destroyed at the end of testing. Significant deviations should be investigated.
10.8.13	***Does the procedure require that personnel involved in administrating the stability program have been properly trained and qualified or have been evaluated to possess the education, training and experience or combination thereof to perform their duties?***	
10.9	**PROCEDURE #9:** Additional Types of Stability Studies	
10.9.1	***Does the procedure address studies designed to obtain more in-depth knowledge of drug substance or drug product?***	

Table 10.1 (Continued)

#	Stability program sub element topic	Notes and clarifications
10.9.2	Does the procedure address the need to perform shipping studies?	Shipping studies are designed to determine the impact of the range of conditions that drug substance or drug product can be subjected to as they are shipped locally or globally.
10.9.3	Does the procedure address the need to perform thermal excursion studies?	To determine the impact of short-term high and low temperature excursions.
10.9.4	Does the procedure address the need to perform bulk hold-time studies?	That is to determine how long and under what conditions finished product can be held before being packaged in its primary container-closure system.
10.9.5	Does the procedure address the need to perform intermediate or blend-hold time studies?	These studies determine how long and under what conditions partially manufactured product such as blends, tablet cores, etc. can be held before they are finally manufactured into finished product.
10.9.6	Does the procedure address the need to perform post application approval change studies?	If changes to the dosage form, manufacturing process, API supplier, manufacturing location, etc. occur, stability studies to assess the impact of these changes to the previously assigned or new shelf-life are most likely necessary.
10.9.7	Does the procedure address or make reference to an additional procedure for collection and storage of retain (e.g. reserve or control) samples for each manufactured batch?	See 21 CFR Sec. 211.170 Reserve samples.

Stability Chambers

Stability chambers are enclosures (e.g. pieces of equipment) that provide stable environmental conditions (e.g. temperature, humidity, and light) for determining and monitoring the physical and chemical stability of APIs and drug products.

Although stability chambers constitute an important and integral part of any stability program for both API and Drug Product, their selection, qualification, use, and maintenance are not even mentioned in any of the international guidances, such as ICH or WHO, let alone the GMPs.

The purpose of the following section is to provide some insight into selection, qualification, use, and maintenance of stability chambers.

Table 10.2 Selection, qualification, calibration, and maintenance of stability chambers.

#	Stability program sub element topic	Notes and clarifications
10.10	**PROCEDURE #10:** Selection, Qualification, Calibration, and Maintenance of Stability Chambers	
10.10.1	***Does the procedure discuss User Requirement Specifications related to stability testing and chamber selection including:***	A User Requirement Specification (URS) is a document that is drawn up by a buyer of equipment to describe precisely the required attributes of the equipment.
	• Selecting the type, make, and model of stability chamber to meet current and future stability testing needs such as: o Reach-In chambers? o Walk-In chambers? o Photostability chambers? o Refrigerated chambers? o Low temperature (freezer) chambers?	This includes both Research and Developmental work in addition to support of regulatory application submission, scale-up, and post approval changes and commercial product stability needs.

Table 10.2 (Continued)

#	Stability program sub element topic	Notes and clarifications
	• Ensuring the stability chamber can operate within the set-points and tolerances of temperature and humidity required by ICH guidelines for all climatic regions, including: ○ ±2 °C for any temperature above refrigerated? ○ ±3 °C for refrigerated? ○ ±5 °C for freezer? ○ ±10 °C for an ultra-low freezer? ○ Relative humidity ±5% RH?	Specifically, as delineated in ICH Guideline Q1A (R2) STABILITY TESTING OF NEW DRUG SUBSTANCES AND PRODUCTS.
	• Ensure that the number of significant figures for these chamber tolerances is specified in the procedure as part of the URS?	For example, if the tolerance is 23–27 °C for a set point of 25 °C, does this mean 22.5 °C, (which rounds to 23 °C) is within specification? If you take the number at face value, without rounding, it will fail the specification range of 23–27 °C.
	• That for any chamber with both temperature and relative humidity component that the specifications are ±2 °C and ±5% RH?	
	• Matching number and size of stability chambers to current and future manufacturing needs?	Potential changes in product manufacturing scales (both increases and decreases) should be taken into account.
	• To determine the re-test requirements, shelf-life, and subsequent expiration dating of API and drug product?	
10.10.2	***Does the procedure address qualification of stability chambers to include:***	
	• Reasons or circumstances required for qualification and re-qualification of stability chambers?	

(Continued)

Table 10.2 (Continued)

#	Stability program sub element topic	Notes and clarifications
	• A requirement that stability chambers are qualified by use of a protocol which includes the following: o Is created, reviewed, approved, and executed by appropriate personnel including Engineering, Testing Laboratory, Stability Program Management, and Quality Assurance personnel? o Has preapproved acceptance criteria at each stage? o Ensures that failures to meet preapproved acceptance criteria are addressed in a timely and appropriate fashion including a risk assessment if actual stability samples may be impacted as in the case for requalification?	
	• Requires that a qualification report is written, reviewed, and approved and includes sections addressing each stage of the qualification?	
	• Contains sections related to **Design Qualification (DQ)**, and are the following points addressed (as appropriate):	Design qualification is an assessment to determine if the chamber's physical and mechanical design is suitable for its intended use and that the equipment manufacturer has implemented appropriate controls for the design, manufacturer, and testing of the chambers.
	o For Commercial Off the Shelf (COTS) stability chambers, that assessments are made to ensure the chambers to be purchased and qualified are suitable for their intended use, including input from Product Development, Engineering, the Laboratory, and QA?	

Table 10.2 (Continued)

#	Stability program sub element topic	Notes and clarifications
	o For Commercial Off the Shelf (COTS) stability chambers, that assessments are made to ensure the chamber equipment manufacturer has utilized appropriate systems for design, manufacture and testing of the chambers which will be purchased?	
	o For customized chamber design that a detailed stand-alone Design Qualification (DQ) was created, approved, and reviewed by appropriate personnel?	
	• Contains sections related to **Installation Qualification (IQ),** and are the following points addressed (as appropriate):	The purpose of IQ is to demonstrate that the chamber was installed according to the manufacturer's specifications, and the user's requirements, as delineated in the protocol.
	o Chamber identification, which is documented in the executed protocol?	
	o Size of the chambers which is documented in the executed protocol?	
	o Chamber location which is documented in the executed protocol?	
	o Description of chamber construction, including drawings, as well as facilities, piping and instrumentation diagrams including water lines and wiring, floor plans, etc. provided by the manufacturer and sourced from the Facilities and Engineering departments, which is documented in the executed protocol?	
	o A spare parts list which will be maintained onsite as recommended by the manufacturer, which is documented in the executed protocol?	

(Continued)

Table 10.2 (Continued)

#	Stability program sub element topic	Notes and clarifications
	o Ongoing preventive maintenance (PM) measures and time tables as recommended by the manufacturer and documented in the executed protocol?	
	o Reference to literature from the manufacturer including requirements for chamber location including environmental conditions which could affect chamber operation such as surrounding temperature, humidity, air circulation space, etc.?	Sections in the executed protocol should address these points and included a "Meets Criteria" statement for each.
	o Electrical requirements?	Include a "Meets Criteria" statement in the protocol for this assessment.
	o Uninterruptable and backup power availability, which is documented in the executed protocol?	
	o Water source and water source quality assessment for ongoing humidity level maintenance including considerations for pH and conductivity as indicated in literature provided by the manufacturer?	Include a "Meets Criteria" statement in the protocol for this assessment.
	o Description of chamber monitoring and alarm systems to be used which is documented in the executed protocol?	
	o Description of chamber failure backup systems which is documented in the executed protocol?	
	o Test operation of equipment including turning system on and whether set-point controls are operational which is documented in the executed protocol?	

Table 10.2 (Continued)

#	Stability program sub element topic	Notes and clarifications
	o Test equipment to be used during IQ and PQ such as thermocouples, humidity meters, pH meters, conductivity meters, etc. which is documented in the executed protocol? o Are applicable chamber-related SOPs such as calibration, maintenance, monitoring, and chamber excursions listed, which is documented in the executed protocol? o Are all deviations from the protocol for IQ, documented and explained which is documented in the executed protocol? o Is the completed IQ reviewed, signed, and shared by and with the appropriate personnel?	
	• Contains sections related to **Operational Qualification (OQ)** and are the following points addressed (as appropriate):	The purpose of OQ is to demonstrate that an empty chamber is operating according to manufacturer's specifications and user's requirements as outlined in the protocol.
	o A statement that confirms the IQ was successfully completed which is documented in the executed protocol? o A statement confirming the chamber turned on (when power was supplied and the on switch activated) and this is recorded in the protocol? o Confirmation that all chamber variables (e.g. °C versus °F, alarm delay, temperature, humidity, etc.) have been identified, set appropriately, and recorded in the executed protocol? o Confirmation the chamber reached intended conditions and stabilize at those conditions and this is recorded in the executed protocol?	

(Continued)

Table 10.2 (Continued)

#	Stability program sub element topic	Notes and clarifications
	o Confirmation the chamber has been calibrated for temperature and meets acceptance criteria?	Include a "Meets Criteria" statement in the protocol for this assessment.
	o Confirmation the chamber has been calibrated for humidity, and meets acceptance criteria?	Include a "Meets Criteria" statement in the protocol for this assessment.
	o Confirmation an empty chamber temperature and humidity mapping (distribution study) has been performed and include the following points:	
	▪ That the purpose of the temperature and humidity mapping study is to demonstrate that the set-point is maintained, with the allowed tolerances, throughout the empty chamber, are established?	
	▪ A description of the temperature and humidity set points?	
	▪ A description of temperature probe location such as: 15 probes placed 5 top, 5 middle, 5 bottom, with one at the geographic center and one probe at each corner or side of the empty shelves?	
	▪ A description of humidity probe location: relative humidity (RH) 15 probes placed in the same position as the temperature probes if possible?	
	▪ A statement that one temperature and humidity probe placed in close proximity to the chamber controller probe?	
	▪ A statement and evidence that the mapping study ran for at least 24 hours?	
	▪ A statement and evidence that the data collection rate was approximately every five minutes?	Five minutes as a minimum.
	▪ A statement that all probes were calibrated prior to use and the documentation for the calibration is included in the OQ section of the qualification report?	

Table 10.2 (Continued)

#	Stability program sub element topic	Notes and clarifications
	▪ Confirmation and evidence that all data points were within specification (±2 °C and ±5% RH of the set point)? ▪ Documentation that all deviations from the protocol have been captured, investigated, and addressed? ▪ Confirmation that the completed OQ section of the protocol has been reviewed, signed, and shared by and with the appropriate personnel?	Include a "Meets Criteria" statement in the protocol for this assessment.
	• Contains sections related to **Performance Qualification (PQ)** and are the following points addressed (as appropriate):	Performance qualification is designed to demonstrate that the chamber, *when it is full of samples*, is performing to the manufacturer's specifications, and the user's requirements, as delineated in the protocol.
	o A statement that confirms OQ was successfully completed, and this is documented in executed protocol? o Confirmation that the temperature and humidity set-points established in the OQ has been unchanged and the chamber is holding those conditions, and this is documented in executed protocol? o Confirmation that a second temperature and humidity mapping (distribution) study has been performed, with *the chamber full* and includes the following points: ▪ That the chamber was filled with empty boxes, containers, or dummy samples and this is documented in executed protocol?	
	▪ That there is a description of how the chamber was filled included in the protocol, and this is documented in executed protocol?	Photographs should be included in the executed protocol to show the loading configuration and types of samples used.

(Continued)

Table 10.2 (Continued)

#	Stability program sub element topic	Notes and clarifications
	▪ That the full-chamber temperature and humidity mapping study was executed in the same fashion as in the empty chamber study performed in the OQ and this is documented in executed protocol?	
	▪ That all data points were within specification and in alignment with the OQ results? (±2 °C and ±5% RH of the set point) and this is documented in executed protocol?	Include a “Meets Criteria” statement in the protocol for this assessment.
	o Confirmation that an open-door study (e.g. leaving a door open to simulate sample placement and removal) was performed and included the following:	
	▪ Test of the maximum time a door could be left open during true operational conditions (e.g. no more than three minutes) and this is documented in executed protocol?	
	▪ That the performance of the chamber open-door study evaluated initiation of the alarm and this is documented in executed protocol?	
	o Confirmation that a simulated chamber failure study was performed and it consider the following:	
	▪ That the failure study was performed at the completion of the full-chamber temperature and humidity mapping (distribution) study and with the probes and empty boxes, or shipping containers, or dummy samples, were still in the same locations and this is documented in executed protocol?	
	▪ That that the purpose of the study, which is to demonstrate the effect of a chamber failure on the chamber conditions and confirm operation of the alarm, is documented in executed protocol?	

Table 10.2 (Continued)

#	Stability program sub element topic	Notes and clarifications
	▪ That the failure study included powering off the chamber once it had achieved its set-point and this is documented in executed protocol?	
	▪ That the failure study determined the time required for the set point to go out of tolerance and this is documented in executed protocol?	
	▪ That the failure study was terminated after four to six hours if it did not go out of specification during this time, and this is documented in executed protocol?	
	▪ That when the chamber went out of specification, the alarm sounded and automatic notification system (if present) was activated, and this is documented in executed protocol?	If chamber does not go out of specification during failure testing, then it should be forced-out to test alarm and notification system.
	o Confirmation that upon completion of the failure study that a recovery study was initiated and considered the following:	
	▪ That the purpose of the recovery study, which was to determine the time required for the chamber to return to within tolerances once the power was restored was addressed and this is documented in executed protocol?	
	▪ That when the power was restored, the time it took for the chamber to return to its set-point was noted and this is documented in executed protocol?	
	o Confirmation that all deviations from the protocol are addressed in writing and this is documented in executed protocol?	

(Continued)

Table 10.2 (Continued)

#	Stability program sub element topic	Notes and clarifications
	o Confirmation that the completed PQ was reviewed, signed, and shared by and with the appropriate personnel and this is documented in executed protocol? o Confirmation that the chamber was placed into service following sign off and approval of all DQ/IQ/OQ/PQ documentation and closure of the change control, which was used to initiate the chamber qualification? • Contains sections related to **stability chamber re-qualification** and are the following points addressed (as appropriate): o When and how stability chamber requalification occurs? o That requalification is executed under a pre-approved protocol in the same manner as during the original qualification? o Confirmation that requalification is undertaken for the following reasons: ▪ The chamber set point has changed? ▪ The chamber has malfunctioned and has been repaired? ▪ A pre-determined operational period, typically one to three years? o Confirmation that requalification consists of the following items: ▪ Repeating the temperature and humidity mapping study performed during PQ but with the contents of the chamber "as is"? ▪ Elimination of the open door, failure, and recover portions of the temperature and humidity mapping study performed during PQ?	

Table 10.2 (Continued)

#	Stability program sub element topic	Notes and clarifications
	▪ If the chamber failed requalification and contains actual stability samples, that a deviation is filed and a thorough investigation and risk assessment is performed to determine the impact on samples and stability study integrity?	
	o Ensuring that all deviations from the protocol are fully documented and explained and captured in the requalification protocol?	
	o Confirmation that the completed requalification documentation is reviewed, signed, and shared by and with the appropriate personnel?	
10.3	***Does the procedure address chamber security and access control and include the following points:***	
	• Physical security and access to the room or area where chambers are located, which may involve:	
	o Secure card entry controls?	
	o Key pad entry controls?	
	o Traditional lock and key entry controls?	
	o Biometric entry controls?	This could include fingerprint scans, hands scan, etc.
	• Physical security and access to the chambers themselves which may involve:	
	o Logbook sign-in, sign-out entry controls?	
	o Secure card entry controls?	
	o Key pad entry controls?	
	o Traditional lock and key entry controls?	
	o Biometric entry controls?	

(Continued)

Table 10.2 (Continued)

#	Stability program sub element topic	Notes and clarifications
	• Limiting entry into stability chambers to only authorized personnel?	This would include stability program management staff, engineering, QA, etc.
	• Ensuring unauthorized personnel who require access are escorted by authorized personnel at all times?	This could include third party maintenance personnel, corporate auditors, visitors, etc.
	• Ensuring that there is a historical, electronic or paper record of chamber entries including date, time and person(s) accessing the chambers?	If appropriate, a record for the reason for entry is also recommended (e.g. stability pull, stability set-down, chamber inventory, etc.)
10.4	***Does the procedure address alarm and monitoring systems and practices and include the following:***	
	• Ensures that appropriate alarm and monitoring systems have been installed, qualified, and maintained, and take the following form:	
	o Manual inspection and recording?	This may include posted inspection sheets with date, time, set conditions, observed conditions and person recording the observations.
	o Paper chart recordings? o Audible alarms? o Computerized system with telephone and e-mail call-out capabilities?	
	• Ensures that chart recorders or manual systems are reviewed at least on a daily basis excluding weekends and holidays?	Some contingency plans may be necessary for extended holidays.
	• If a computerized system with e-mail and call-out functionality is installed, then has it been properly validated during chamber qualification?	
	• If a computerized system with e-mail and call-out functionality is installed then personnel who need to be contacted by the system are aware of their responsibilities, the list is current and the personnel on the list are receiving notifications as necessary?	Primary contacts should have backup contacts. The procedure also needs to take into account notification of backup personnel when primaries are out on vacation days, holidays or sick days.

Table 10.2 (Continued)

#	Stability program sub element topic	Notes and clarifications
10.5	***Does the procedure discuss the impact and actions required when temperature and humidity excursions occur including the following points:***	Excursions are when temperature and/or humidity values go outside their validated set-points and ranges.
	• A requirement to address excursions that exceed specified set points and ranges for more than 24 hours?	Excursions of less than 24 hours, generally do not need to be addressed, unless the excursion is extreme and impacts the kinetic mean temperature. Ranges are typically ±2 °C and/or ±5% RH.
	• Categorize the type of failure such as mechanical failure, power failure, personnel error, etc.?	
	• Assessment of the impact of the failure?	Chamber excursions of greater than 24 hours require opening of a deviation report.
	• Suggestions for addressing failures including:	
	o Moving samples to a back-up chamber?	
	o Pulling and analyzing, some samples for physical and/or chemical analysis to assess the impact of the chamber excursion or failure?	In cases where the failure was extreme and placed the samples at risk of degradation, this interventional testing may be appropriate.
	o Adding more time to the study?	This may be appropriate if the excursion or failure places the samples in a less severe condition than was defined in the stability protocol.
	• Offers guidance on completing the formal excursion report, including:	
	o Determining the impact on stored samples and study integrity?	
	o Regulatory agency notification?	
	o Chamber repair?	
	o The need for chamber requalification?	
	o Completion of OOS or deviation investigations?	

(Continued)

Table 10.2 (Continued)

#	Stability program sub element topic	Notes and clarifications
	• Establishing and maintaining redundant systems to prevent or mitigate the impact of stability chamber failures to include: o Installing and maintaining backup power systems? o Alternate water systems? o Alternate chambers?	
	• The existence of a disaster recovery plan which addresses stability chambers and stability samples, which might include: o Shipping samples to an alternate company location? o Shipping samples to a contract stability storage and/or testing facility?	Disaster recovery plans are global in nature but need to consider the impact of the disaster on samples which are under stability studies.
10.6	***Does the procedure discuss periodic chamber calibration and preventive maintenance requirements including:***	
	• During calibration, to place certified temperature and humidity measuring devices inside the chamber near the probe used by the chamber to control temperature or humidity?	
	• Establishing acceptance limits for calibration of chamber settings?	For calibration it is typically $\pm 1\,^{\circ}C$ and $\pm 3\%$ RH.
	• Establishing a calibration schedule with assurances it is being followed?	Stability chamber calibration is typically on a 6-month schedule.
	• Addressing how to handle stability chamber calibration failures? • Stability chamber preventive maintenance (PM) including: o Cleaning heating and cooling coils?	
	o Cleaning the humidity reservoir?	The reservoir catches condensation which may collect in the system.
	o Checking and cleaning stability chamber door seals? o Clean chamber surfaces? o Other tasks as recommended by the chamber manufacturer?	

Table 10.2 (Continued)

#	Stability program sub element topic	Notes and clarifications
10.7	***Does the procedure address the need for periodic chamber sample inventories and reconciliation and include the following:***	
	• Provides guidance on the design and implementation of a sample logging and tracking system, which may be manual or electronic?	Electronic systems can include • Manual input into database. • Barcode scanning when adding or removing samples.
	• Requires periodic chamber inventory and reconciliation of records to determine chamber contents?	
	• Requires physical chamber inventory to confirm chamber contents versus chamber sample logs or electronic database? o Random statistical sample selection for large chambers? o Full chamber inventories for smaller chambers?	Inventories are required at least annually.

Tools and Templates

The following are provided in electronic format as Example Tools and Templates, which can be found in the Chapter 10 Appendix (www.wiley.com/go/Bliesner/LabControl_GMPEnvironment):

- Topics in Table 10.1 in checklist format for use in evaluating, upgrading, or establishing your own laboratory stability program sub element
- Topics in Table 10.2 in checklist format for use in evaluating, upgrading, or establishing your own laboratory stability program, stability chambers, sub element
- Example-Template Stability Protocol
- Example-Template Stability Report (Matrix Format)
- Model Stability Program-List of Example Standard Operating Procedures

Glossary[1]

Accelerated Testing Studies designed to increase the rate of chemical degradation or physical change of a drug substance or drug product by using exaggerated storage conditions as part of the formal stability studies. Data from these studies, in addition to long-term stability studies, can be used to assess longer-term chemical effects at non-accelerated conditions and to evaluate the effect of short-term excursions outside the label storage conditions such as might occur during shipping. Results from accelerated testing studies are not always predictive of physical changes.

Bracketing The design of a stability schedule such that only samples on the extremes of certain design factors, e.g. strength and package size, are tested at all time points as in a full design. The design assumes that the stability of any intermediate level is represented by the stability of the extremes tested. Where a range of strengths is to *Stability Testing of New Drug Substances and Products* be tested, bracketing is applicable if the strengths are identical or very closely related in composition (e.g. for a tablet range made with different compression weights of a similar basic granulation, or a capsule range made by filling different plug fill weights of the same basic composition into different size capsule shells). Bracketing can be applied to different container sizes or different fills in the same container closure system.

Climatic Zones The four zones in the world that are distinguished by their characteristic prevalent annual climatic conditions. This is based on the concept described by W. Grimm (*Drugs Made in Germany*, 28:196–202, 1985 and 29:39–47, 1986).

1 Adapted from the ICH guideline Q1A R2 listed in the references earlier.

Commitment Batches	Production batches of a drug substance or drug product for which the stability studies are initiated or completed post approval through a commitment made in the registration application.
Container-Closure System	The sum of packaging components that together contain and protect the dosage form. This includes primary packaging components and secondary packaging components, if the latter are intended to provide additional protection to the drug product. A packaging system is equivalent to a container closure system.
Dosage Form	A pharmaceutical product type (e.g. tablet, capsule, solution, cream) that contains a drug substance generally, but not necessarily, in association with excipients.
Drug Product	The dosage form in the final immediate packaging intended for marketing.
Drug Substance	The unformulated drug substance that may subsequently be formulated with excipients to produce the dosage form. Also referred to as Active Pharmaceutical Ingredient or API.
Excipient	Anything other than the drug substance in the dosage form.
Expiration Date	The date placed on the container label of a drug product designating the time prior to which a batch of the product is expected to remain within the approved shelf-life specification if stored under defined conditions, and after which it must not be used. *Stability Testing of New Drug Substances and Products.*
Formal Stability Studies	Long term and accelerated (and intermediate) studies undertaken on primary and/or commitment batches according to a prescribed stability protocol to establish or confirm the re-test period of a drug substance or the shelf-life of a drug product.
Impermeable Containers	Containers that provide a permanent barrier to the passage of gases or solvents, e.g. sealed aluminum tubes for semi-solids, sealed glass ampoules for solutions.

Intermediate Testing Studies conducted at 30 °C/65% RH and designed to moderately increase the rate of chemical degradation or physical changes for a drug substance or drug product intended to be stored long term at 25 °C.

Long-Term Testing Stability studies under the recommended storage condition for the re-test period or shelf-life proposed (or approved) for labeling.

Mass Balance The process of adding together the assay value and levels of degradation products to see how closely these add up to 100% of the initial value, with due consideration of the margin of analytical error.

Matrixing The design of a stability schedule such that a selected subset of the total number of possible samples for all factor combinations is tested at a specified time point. At a subsequent time point, another subset of samples for all factor combinations is tested. The design assumes that the stability of each subset of samples tested represents the stability of all samples at a given time point. The differences in the samples for the same drug product should be identified as, for example, covering different batches, different strengths, different sizes of the same container closure system, and, possibly in some cases, different container closure systems.

Mean Kinetic Temperature A single derived temperature that, if maintained over a defined period of time, affords the same thermal challenge to a drug substance or drug product as would be experienced over a range of both higher and lower temperatures for an equivalent defined period. The mean kinetic temperature is higher than the arithmetic mean temperature and takes into account the Arrhenius equation.
When establishing the mean kinetic temperature for a defined period, the formula of J. D. Haynes (*J. Pharm. Sci.*, 60:927–929, 1971) can be used.

New Molecular Entity An active pharmaceutical substance not previously contained in any drug product registered with the national or regional authority concerned. A new salt, ester, or non-covalent-bond derivative of an approved drug substance is considered a new molecular entity for the purpose of stability testing under this guidance. *Stability Testing of New Drug Substances and Products.*

Pilot Scale Batch A batch of a drug substance or drug product manufactured by a procedure fully representative of and simulating that to be applied to a full production scale batch. For solid oral dosage forms, a pilot scale is generally, at a minimum, one-tenth that of a full production scale or 100,000 tablets or capsules, whichever is the larger.

Primary Batch A batch of a drug substance or drug product used in a formal stability study, from which stability data are submitted in a registration application for the purpose of establishing a re-test period or shelf-life, respectively. A primary batch of a drug substance should be at least a pilot scale batch. For a drug product, two of the three batches should be at least pilot scale batch, and the third batch can be smaller if it is representative with regard to the critical manufacturing steps. However, a primary batch may be a production batch.

Production Batch A batch of a drug substance or drug product manufactured at production scale by using production equipment in a production facility as specified in the application.

Re-Test Date The date after which samples of the drug substance (API) should be examined to ensure that the material is still in compliance with the specification and thus suitable for use in the manufacture of a given drug product.

Re-Test Period	The period of time during which the drug substance is expected to remain within its specification and, therefore, can be used in the manufacture of a given drug product, provided that the drug substance has been stored under the defined conditions. After this period, a batch of drug substance destined for use in the manufacture of a drug product should be re-tested for compliance with the specification and then used immediately. A batch of drug substance can be re-tested multiple times and a different portion of the batch used after each re-test, as long as it continues to comply with the specification. For most biotechnological/biological substances known to be labile, it is more appropriate to establish a shelf-life than a re-test period. The same may be true for certain antibiotics.
Semi-Permeable Containers	Containers that allow the passage of solvent, usually water, while preventing solute loss. The mechanism for solvent transport occurs by absorption into one container surface, diffusion through the bulk of the container material, and desorption from the other surface. Transport is driven by a partial-pressure gradient. Examples of semi-permeable containers include plastic bags and semi-rigid, low-density polyethylene (LDPE) pouches for large volume parenterals (LVPs), and LDPE ampoules, bottles, and vials.
Shelf-Life (Also Referred to As Expiration Dating Period)	The time period during which a drug product is expected to remain within the approved shelf-life specification, provided that it is stored under the conditions defined on the container label.
Specification: Release	The combination of physical, chemical, biological, and microbiological tests and acceptance criteria that determine the suitability of a drug product at the time of its release.

Specification: Shelf-Life	The combination of physical, chemical, biological, and microbiological tests and acceptance criteria that determine the suitability of a drug substance throughout its re-test period, or that a drug product should meet throughout its shelf-life.
Stability Chamber	Stability chambers are enclosures (e.g. pieces of equipment) that provide stable environmental conditions (e.g. temperature, humidity, and light) for determining and monitoring the physical and chemical stability of APIs and drug products. These are also referred to in some cases as environmental chambers.
Storage Condition Tolerances	The acceptable variations in temperature and relative humidity of storage facilities for formal stability studies. The equipment should be capable of controlling the storage condition within the ranges defined in this guideline. The actual temperature and humidity (when controlled) should be monitored during stability storage. Short-term spikes due to opening of doors of the storage facility are accepted as unavoidable. The effect of excursions due to equipment failure should be addressed and reported if judged to affect stability results. Excursions that exceed the defined tolerances for more than 24 hours should be described in the study report and their effect assessed.
Stress Testing (Drug Substance)	Studies undertaken to elucidate the intrinsic stability of the drug substance. Such testing is part of the development strategy and is normally carried out under more severe conditions than those used for accelerated testing.
Stress Testing (Drug Product)	Studies undertaken to assess the effect of severe conditions on the drug product. Such studies include photostability testing (see ICH Q1B) and specific testing on certain products, (e.g. metered dose inhalers, creams, emulsions, refrigerated aqueous liquid products).

Supporting Data Data, other than those from formal stability studies, that support the analytical procedures, the proposed re-test period or shelf-life, and the label storage statements. Such data include (i) stability data on early synthetic route batches of drug substance, small scale batches of materials, investigational formulations not proposed for marketing, related formulations, and product presented in containers and closures other than those proposed for marketing; (ii) information regarding test results on containers; and (iii) other scientific rationales.

References

1 21 CFR Part 211, Section 211.166 Stability Testing.
2 ICH Q1A R2 Guideline STABILITY TESTING OF NEW DRUG SUBSTANCES AND PRODUCTS.
3 ICH Q1E Guideline EVALUATION FOR STABILITY DATA.
4 Kim, H.-B. (ed.) (2010). *Handbook of Stability Testing in Pharmaceutical Development*. Springer Science + Business Media, LLC. ISBN: 978-1-4419-2756-9.
5 D.M. Bliesner "*Establishing a CGMP Laboratory Audit System*", Wiley-Interscience 2006, ISBN-10: 0-471-73840-9
6 Kim, H.-B. (2010). *Pharmaceutical Stability Testing to Support Global Markets*. Springer Science + Business Media, LLC. ISBN: 978-1-4419-0888-9.
7 ICH Q7 Guideline for GOOD MANUFACTURING PRACTICE GUIDE FOR ACTIVE PHARMACEUTICAL INGREDIENTS.
8 World Health Organization (WHO) (2009). Annex 2 Stability Testing of Active Pharmaceutical Ingredients and Finished Pharmaceutical Products. *Technical Report Series, No.* 953.
9 21 CFR Sec. 314.81 Other post marketing reports. (Field Alert Reporting).

11

Components of the General Laboratory Compliance Practices Sub Element (CP)

Description of the General Laboratory Compliance Practices Sub Element

General laboratory compliance practices are those day-to-day practices and activities, which everyone working in a Current Good Manufacturing Practice (CGMP) compliant laboratory are expected to know and understand, yet they might not be clearly described or codified in writing.

These practices and activities are listed separately in this chapter because they may not easily fit into the sub elements previously described. However, they may be somewhat redundant to items identified in the Managerial and Administrative System sub element. It should be noted that these compliance practices and activities may be considered "soft skills" and have more to do with the implementation and evaluation of the effectiveness of the previously described sub elements.

The General Laboratory Compliance Practices sub element includes at least 13 individual topics. A laboratory, which is in compliance with CGMPs, should address each of these topics (as applicable) even though there are no strict requirements in the CGMP regulations themselves, nor any formal, documented industry guidance.

These topics may include the following: (i) Housekeeping and General Appearance, (ii) Sample Handling and Storage, (iii) Standard Handling and Storage, (iv) Reagent Handling and Storage, (v) Instruments and Equipment, (vi) Laboratory Water and Solvents and Solutions, (vii) Laboratory Facilities and Services, (viii) Laboratory Supplies, (ix) Laboratory Glassware, (x) Analyst Documentation Practices, (xi) Laboratory Safety – High Level Review, (xii) Analysts Demeanor and Behavior, and (xiii) Supervisor Presence and Effectiveness.

Laboratory Control System Operations in a GMP Environment, First Edition. David M. Bliesner.

Companion website: www.wiley.com/go/Bliesner/LabControl_GMPEnvironment

Contents of the Sub Element

The 13 topics previously mentioned are presented in Table 11.1 in the form of questions related to each topic. For each topic, additional detail or "points to consider" are offered along with notes and clarifications as appropriate. The questions, notes, and clarifications are all based on actual inspectional findings from Regulatory Inspectors or points, which have been found to directly or indirectly impact compliant and efficient real-world laboratory control system operations.
It is envisioned that this table may be used as a basis for:

- Daily walk-through or scheduled periodic self-inspections of the laboratory
- A means to organize, capture, track, and pattern challenges or gaps related to general laboratory compliance practices
- As a basis or integration into the corrective and preventive action (CAPA) system
- As an augmentation to periodic laboratory CGMP training
- As an augmentation to periodic laboratory efficiency assessments

Table 11.1 Components of the general laboratory compliance practices sub element.

#	General laboratory compliance practices sub element topic	Notes and clarifications
11.1	***Housekeeping and General Appearance***	How a laboratory "shows" can have a substantial impact on the level of scrutiny with which a Regulatory Inspector will take. If the initial impression is that the lab is in a physical state of disarray, Inspectors immediately extrapolate that compliance challenges probably exist as well. In this circumstance, Inspectors are more likely to spend time in the lab to confirm their suspicions.
11.1.1	Is there sufficient laboratory space for analyst to perform their expected duties?	Section 211.42 of the Good Manufacturing Practices (GMPs) titled "Design and construction features" states that there shall be "…adequate space for the orderly placement of equipment…" applies to laboratory areas as well as manufacturing or warehousing spaces. A crowded laboratory suggests a lack of commitment to quality at the upper levels within the organization.
11.1.2	Is the laboratory generally clean?	
11.1.2.1	Is this true for the entire lab?	

Table 11.1 (Continued)

#	General laboratory compliance practices sub element topic	Notes and clarifications
11.1.3	Are bench tops where analysts are working generally neat and well-organized leading to efficient and CGMP compliant work practices?	
11.1.4	Are bench tops where analysts are not working clean and uncluttered?	Benchtops, where there are old samples and sample solutions, become compliance flags and will attracted unwanted scrutiny during an inspection. This is particularly true with respect to Data Integrity challenges. Unsecured samples and sample solutions connote the ability to prepare alternant, backup, or trial samples. The practice of injecting "trial samples" (e.g. screening of samples to see if they will meet specifications prior to their analysis) has become one of the first things US Food and Drug Administration (FDA) inspectors look for when reviewing chromatographic records. Numerous organizations have been cited for this practice, as review of FDA Warning Letters will reveal this to be true.
11.1.5	Are sinks clear of dirty glassware, spatulas, stir bars, or other items?	
11.1.6	Are shelves within the laboratory and surrounding areas clean and uncluttered?	
11.1.7	Are the adjacent areas to the laboratory listed below clean and uncluttered and free of uncontrolled samples, records, logbooks, and notebooks, etc.:	See 11.1.3.
11.1.7.1	Common areas such as break rooms?	
11.1.7.2	Workspaces and cubicles?	Stacks of notebooks, either old or new, are a compliance flag. It implies a lack of control in the laboratory documentation system.

(Continued)

Table 11.1 (Continued)

#	General laboratory compliance practices sub element topic	Notes and clarifications
11.1.7.3	Offices?	
11.1.7.4	Reference collections such as literature, manuals, etc.?	Equipment and instrument manuals should be controlled, yet accessible to laboratory personnel as needed.
11.1.8	Are there separate waste receptacles being used for office waste (e.g. paper and packaging), laboratory waste (e.g. broken or disposable glassware and plastic ware), and chemical waste (e.g. HPLC, dry chemical, solutions, etc.)?	
11.2	***Sample Handling and Storage***	See Warning Letter WL: 320-13-21, 18 July 2013 for examples of US FDA findings related to sample handling.
11.2.1	Are received samples logged into a controlled sample receipt system?	Due to widespread data integrity challenges, expectations with respect to sample handling and tracking have become more rigorous. Sample chain of custody is no longer a concern for only controlled substances, but samples in general.
11.2.2	Are all samples for testing which are delivered to the laboratory stored appropriately, including:	
11.2.2.1	Protected from light?	As necessary.
11.2.2.2	Stored in a similar manner to controlled substances?	The level of control required for controlled substances is greater, however all samples need to be controlled in a chain-of-custody-like manner.
11.2.2.3	Stored in a refrigerator or freezer?	As necessary.
11.2.2.4	Stored in desiccators?	As necessary.
11.2.3	Are samples on-test placed in a controlled area, under appropriate storage conditions?	

Table 11.1 (Continued)

#	General laboratory compliance practices sub element topic	Notes and clarifications
11.2.4	Are samples retained and stored properly following analyses so that in the event of a laboratory investigation they can be scrutinized or re-tested if necessary and appropriate?	Samples should not be disposed of until the data have been reviewed and the material or product has been released by quality assurance (QA).
11.2.5	Are samples used for training analysts segregated from actually finished product and active ingredient samples?	Training samples should not be collected from products which are in or will be released to market. If an analyst obtains as OOS result during training, and the product lot from which the sample was taken is still in the market, a formal OOS investigation may be warranted. Consider using expired, prepared or composited samples for training.
11.3	***Standard Handling and Storage***	
11.3.1	Are purchased reference standards being stored according to their label or certificate of analysis instructions?	
11.3.2	Are purchased standards being used within expiry?	
11.3.3	Are expired purchased standards disposed in a timely manner to prevent continued use (e.g. Such as USP standards which should be reconciled against the USP-NF)?	
11.3.4	Is the preparation of standard solutions properly recorded in a logbook or other record keeping system?	
11.3.5	Are prepared standard solutions properly labeled including contents, concentration, date prepared, initials of preparer, special storage conditions, and expiration date?	This is a typical example of an enforceable compliance practice in the laboratory, which isn't described in the Regulations.

(Continued)

Table 11.1 (Continued)

#	General laboratory compliance practices sub element topic	Notes and clarifications
11.3.6	Are prepared standard solutions being stored according to the supporting validation data?	One of the components in method validation is the determination of standard solution stability. This information should be contained in the final working test procedures.
11.3.7	Are prepared standard solutions being disposed of at the end of the expiry?	
11.4	***Reagent Handling and Storage***	
11.4.1	Are all reagents properly labeled including receipt date, opened data, expiration date, and the initials of the receiver and person opening the container?	
11.4.2	Are all reagents being used within the assigned expiration date as delineated in an SOP?	All reagents need to be assigned an expiration date. This includes materials which are generally expected to be stable (e.g. salts or solvents). The expiration date is determined experimentally, through literature research or by documentation obtained from the supplier.
11.4.3	Are all reagents being stored appropriately including segregation by reactivity and controlled by some type of inventory system?	
11.5	***Instruments and Equipment***	
11.5.1	Are analysts recording equipment use in associated equipment use logs?	
11.5.1.1	Are usage log entries completed at the time of equipment or instrument use?	

Table 11.1 (Continued)

#	General laboratory compliance practices sub element topic	Notes and clarifications
11.5.1.2	Are usage log entries of sufficient detail so that they may be used in the future to conduct laboratory investigations should the need arise?	Details should include at a minimum: the equipment being used, calibration or qualification status, the person performing the analysis, samples being analyzed, and any observations related to equipment function/malfunction during the analysis.
11.5.2	Are instruments and equipment immediately tagged-out and locked-out via a formal, documented system, when their performance is suspect or when they have past their maintenance, calibration, or requalification time points?	
11.5.2.1	Are issues which lead to tagged-out instruments and equipment which include repairs, calibrations and preventative maintenance being performed within a reasonable time frame?	Organizations with compliance challenges are sometimes observed to have instruments and equipment, which have been placed out of service for extended periods of time. Likewise, new instruments still in the original packaging awaiting IQ/OQ/PQ for extended periods of time are also a compliance flag.
11.5.3	Are daily balance checks recorded in balance usage logs or in logs specifically designed to record daily usage checks?	Proper performance and recording of daily (or time of use) balance checks can be critical. Since weighing is a fundamental part of sample analysis, finding a balance which is out of calibration and has not been calibrated in some time raises concern about all the previous analyses and the corresponding product, which was released to the market. If inaccurate records were kept, and/or a calibration failure occurs, an extensive investigation may be required to ensure that product in the market place meets the quality standards purported on the label.

(Continued)

Table 11.1 (Continued)

#	General laboratory compliance practices sub element topic	Notes and clarifications
11.5.4	Are equipment and instrument usage logs being reviewed by supervisors or quality personnel to ensure that potential failures of equipment do not impact test results?	
11.5.5	Are instruments and equipment left in a useable condition for the next analyst, in particular:	
11.5.5.1	Balances and weighing areas free of spilled reagents and chemicals?	
11.5.5.2	HPLC columns and systems have been properly rinsed and subsequently stored in a suitable solution?	HPLC column care and use can have a substantial impact on sample analysis. Improper column care can actually change the retention characteristics of the column, which can result in changes in the chromatography for the analysis.
11.5.5.3	Extraneous materials have been removed from glove boxes or isolators?	
11.5.5.4	pH meter electrodes stored in proper solutions, and the solutions are not expired?	pH is another fundamental measurement in the laboratory which can be placed at risk if electrodes are not stored properly. Improper storage can modify or destroy the electrode altogether.
11.5.5.5	Auto titrators left clean and ready for next analysis?	
11.5.5.6	Areas, materials, and solutions used for manual titrations left clean and properly stored to prevent cross-contamination or degradation of reagents?	

Table 11.1 (Continued)

#	General laboratory compliance practices sub element topic	Notes and clarifications
11.5.5.7	GC columns and systems have been left in a proper state of storage and ready for next analysis?	Same care as HPLC columns should be applied. In many QC laboratories GC is a less commonly used instrument. Therefore, ensure that the personnel using, maintaining and calibrating the equipment and supplies have the proper level of training and experience related to gas chromatography. If not, it may be appropriate to source expertise from the equipment vendor or an organization specializing it GC maintenance, calibration, and supplies.
11.5.5.8	Absorption and emission spectrometers have been left in a proper state of storage and ready for next analysis?	
11.6	***Laboratory Water and Solvents and Solutions***	
11.6.1	Are the types of water being used in the laboratory appropriate for their intended use as defined by the current USP?	There are several different types and specifications of water listed in the USP <1231> WATER FOR PHARMACEUTICAL PURPOSES.
11.6.2	Has the laboratory water system or systems been properly validated and is there a written water quality monitoring program in place?	If laboratory water is sourced from the manufacturing area, it is important to consider any impact maintenance of that system may have on the quality of laboratory water. The QC laboratory should be part of any change control procedures related to the water system.
11.6.3	Has an individual or individuals been assigned the responsibilities for monitoring water quality and is that task be performed on a routine basis?	
11.6.4	Is water which is taken from the laboratory water system being stored, labeled, and assigned an appropriate expiration date (e.g. water stored in carboys and rinse bottles)?	Pure HPLC grade or reagent grade water, when exposed to the open environment, will begin growing microbes within several hours. The maximum time that water should be stored in the lab is 24 hours. The presence of bacteria in laboratory water can lead to extraneous peaks in chromatographic systems.

(Continued)

Table 11.1 (Continued)

#	General laboratory compliance practices sub element topic	Notes and clarifications
11.6.5	Is purchased bottled water being assigned an appropriate expiration date following opening?	Again, 24 hours at max, unless a stability study provides data to the contrary.
11.6.6	Are mobile phases and other prepared solutions properly labeled and assigned expiration dates?	Expiration dates need to be determined by collecting the appropriate supporting data. This is most often collected during method validation.
11.6.7	Are expired mobile phases and other prepared solutions removed from the laboratory and properly disposed when they expire?	Expired mobile phases can cause changes in chromatography and in some circumstances generate unknown or extraneous peaks in chromatograms.
11.6.8	Are solvents and water typically placed in "squirt bottles" labeled with appropriate expirations dates?	
11.7	***Laboratory Facilities and Services***	
11.7.1	Is the temperature and humidity in the laboratory controlled and appropriate for laboratory testing?	Temperature and humidity control, particularly in an older facility can be significantly challenging. In such circumstances, ensure that the methods used during analysis can actually be performed as written. For example, if your solvents freeze at ambient temperature (which can happen with acetic acid or dimethyl sulfoxide) it may not be possible to pipette the material.
11.7.2	Are instruments critical to testing attached to uninterruptable power supplies?	
11.7.3	Are the following services or systems available and operational without interruptions:	
11.7.3.1	Chromatographic data acquisitions and processing stations?	
11.7.3.2	Intranet and internet access as appropriate?	

Table 11.1 (Continued)

#	General laboratory compliance practices sub element topic	Notes and clarifications
11.7.3.3	Laboratory information management systems (LIMS)?	
11.7.3.4	Local area network and shared drive access?	
11.7.3.5	Networked printers?	
11.7.3.6	Online document control systems?	
11.7.3.7	Work flow scheduling and tracking systems?	
11.7.3.8	Other services?	
11.8	***Laboratory Supplies***	
11.8.1	Is there a chemical inventory list and is it kept current?	
11.8.2	Are appropriate and necessary quantities of chemicals and reagents kept on hand by monitoring the chemical inventory list?	Individuals should be assigned to these duties. If the laboratory is large enough this may require a full-time technician.
11.8.3	Is there a formal ordering procedure for laboratory chemicals, reagents, and supplies which the analysts know how to use?	Ensure that purchasing personnel cannot "swap out" reagents and other supplies "at will." This can result in unintended consequences. For example, using tetrahydrofuran (THF), which has a preservative for an application that requires preservative free THF can result in OOS events. Additional examples include switching out of syringes or filters whose appropriateness for use was not determined during method validation can lead to modification of samples or generate extraneous peaks in chromatograms.
11.8.4	Does the laboratory have a stockroom which houses necessary supplies and is there an individual or individuals responsible for its operation?	See comments in 11.8.2.

(Continued)

Table 11.1 (Continued)

#	General laboratory compliance practices sub element topic	Notes and clarifications
11.9	***Laboratory Glassware***	
11.9.1	Is laboratory glassware which contains sample and standard solutions properly labeled to include contents, concentration, date prepared, expiration date, test method used, and individual who prepared solutions?	Not explicitly cited in any regulation, but often ends up an observation in Warning Letters.
11.9.2	Is clean glassware segregated from dirty glassware?	
11.9.3	Is laboratory glassware washed using a validated process (manual or mechanical)?	Dirty glassware is notoriously problematic and often leads to investigations related to unknown chromatographic peaks. Validation (either manual or mechanical) should be performed, at a minimum, on glassware which is dirtied with a sample or standard which leaves the most challenging residue. Laboratory personnel must be trained to determine whether the glassware they are using is clean or dirty and do not use dirty or suspect glassware.
11.9.4	Do analysts routinely check if glassware is cracked, chipped, scratched, or excessively worn prior to use?	Dispose all glassware which is suspect. The cost of new glassware is much less than the cost of a single laboratory investigation.
11.9.5	Is there visible evidence that analysts know the proper ways to use volumetric glassware such as pipettes, volumetric flasks, and burettes?	For example, if volumetric glassware is dried at high temperature by analysts, it may no longer be within tolerance.
11.9.6	Is there evidence to suggest that analysts know how to determine whether laboratory glassware is properly cleaned?	Solvents sheet cleanly off the side of the glassware leaving no clumps or puddles.

Table 11.1 (Continued)

#	General laboratory compliance practices sub element topic	Notes and clarifications
11.10	***Analyst Documentation Practices***	
11.10.1	Are testing records, notebooks, record books, batch records, etc. controlled in an appropriate fashion before, during, and after testing?	
11.10.2	Are analysts using sticky notes, scratch paper, calendars, or other places to record raw data, make notes, or perform calculations?	This is a chronic problem, which is still observed in many modern laboratories. If you think this is not the case in your own laboratory, periodically perform a search of waste receptacles. You may be surprised.
11.10.3	Are analysts documenting events, observations and calculations in real-time (e.g. contemporaneously) as they perform analyses?	This is a chronic problem in the laboratory and the manufacturing areas where paper-based systems are still in use.
11.10.4	Do analysts use proper documentation error correction procedures such as crossing-out mistakes, dating, explaining, and initialing the corrections?	These procedures need to be codified in an SOP and all personnel trained and periodically re-trained on the procedure.
11.10.5	Is there a rounding and reporting SOP and is it being followed by laboratory analysts?	
11.10.6	Are analysts aware of the definition of raw data and the mandate to maintain the integrity of such data?	See Chapter 9: Components of the Laboratory Data Governance and Data Integrity Sub Element for details.
11.10.7	Are data capture instruments such as hard-bound notebooks, controlled data sheets, preparation logs, logbooks, LIMS entries, or electronic notebooks used in the laboratory?	

(Continued)

Table 11.1 (Continued)

#	General laboratory compliance practices sub element topic	Notes and clarifications
11.10.7.1	Is their issuance, use, review, retrieval, and archival controlled by an SOP?	
11.10.7.2	Is the notebook or logbook SOP followed?	
11.11	***Laboratory Safety – High Level Review***	
11.11.1	Is there evidence that safety in the laboratory is taken seriously (e.g. evidence of use of personal protective equipment, no food or drink in the lab, etc.)?	Safety is not a CGMP requirement and will not normally be addressed by Regulators. However, if an organization does not enforce basic laboratory safety practices, it's pretty much a guarantee that will have CGMP compliance challenges as well.
11.11.2	Does clutter pose unsafe circumstances (e.g. materials stacked on top of benches, samples and reagents with chemical incompatibilities stored together, reagents, and chemicals stored overhead and difficult to reach by shorter analysts, etc.)?	
11.11.3	Are fume hoods used primarily for analyses and not for storage of chemical waste, reagents, and other components?	Smoke studies show that laboratory hood airflow is easily disrupted by only a few objects placed in a hood.
11.11.4	Is there evidence of water or solvent spills on the floor of the laboratory?	
11.11.5	Are solvent waste, HPLC waste, and other laboratory waste containers emptied regularly?	
11.11.6	Are common areas, offices spaces, or cubicles where food storage and consumption are allowed properly separated from testing and chemical storage areas?	

Table 11.1 (Continued)

#	General laboratory compliance practices sub element topic	Notes and clarifications
11.12	***Analysts Demeanor and Behavior***	This is not a CGMP category per say. However, Inspectors are quick to tune into inappropriate attitudes and behavior, which signals them to perform more thorough inspections.
11.12.1	Is there evidence that analysts are using written procedures during testing?	Expectations are that test methods will be used during testing and the associated standard operating procedures (SOPs) be readily available for access and reference.
11.12.2	Do analysts display a positive attitude when questioned about their work or laboratory practices?	Personnel who are poorly trained, over-worked, and under-appreciated will display less than positive attitudes. Additionally, such personnel are less likely to comply with the GMPs.
11.12.3	Do analysts "push the envelope" with respect to use of personal protective equipment such as safety glasses?	
11.12.4	Is there evidence of analyst frustration with respect to working conditions, workload, or availability of resources?	
11.12.5	Are analysts held accountable for the quality of their work including generation of Out-of-Specification (OOS) results?	Whenever possible, the analyst generating the OOS should be involved in the investigation is some fashion. Nothing quite like suffering from your own mistakes to increase compliance with GMPs.
11.13	***Supervisor Presence and Effectiveness***	
11.13.1	Do supervisors know their analysts by sight and name?	
11.13.2	Do supervisors interact directly with their subordinates during the work shift?	Lack of supervision at the bench has become a challenge in the industry. Supervisor should visit the laboratory as many times as practical during their work hours.
11.13.3	Are after hours work shifts properly staffed with supervisors?	Lack of supervision and lack of analytical expertise is a common problem during overnight shift work.

(Continued)

Table 11.1 (Continued)

#	General laboratory compliance practices sub element topic	Notes and clarifications
11.13.4	Do analysts seek out their supervisors when they need assistance or guidance?	
11.13.5	Do supervisors feel they have the proper resources at their disposal to satisfactorily execute their job functions?	
11.13.6	Do supervisors and managers feel they have proper support and the necessary resources from their management to operate the laboratory in a CGMP compliant fashion?	

Tools and Templates

The following are provided in electronic format as Example Tools and Templates, which can be found in the Chapter 11 Appendix (www.wiley.com/go/Bliesner/LabControl_GMPEnvironment):

- Topics in Table 11.1 in checklist format for use during self-inspections.
- Example-Templates of laboratory self-inspection observation capture forms:
  - Completed example
  - Template for implementation

12

Summary for Establishing and Maintaining a Laboratory Control System

A Brief Review of the Laboratory Control System and Its Sub Elements

This chapter presents a brief review of the main topics covered in this text by re-introducing and describing the 10 Laboratory Control System (LCS) sub elements. In addition, some examples of how organizations have failed to properly implement the sub elements will be presented in the form of excerpts from Warning Letter Citations and Form 483 Observations. The Current Good Manufacturing Practices (CGMPs) themselves will also be referenced to provide clarification where appropriate.

In addition, the author offers some final thoughts on establishing and maintaining a CGMP compliant LCS.

How Things Can Go Wrong: Examples of Some Regulatory Citations Organized by Sub Element

Table 12.1 shows only a few examples of how some organizations have gotten into trouble with US Food and Drug Administration (FDA). The reader is encouraged access the following links for additional examples and information related to FDA compliance actions:

- https://www.fda.gov/inspections-compliance-enforcement-and-criminal-investigations/compliance-actions-and-activities/warning-letters
- https://datadashboard.fda.gov/ora/

Laboratory Control System Operations in a GMP Environment, First Edition. David M. Bliesner.

Companion website: www.wiley.com/go/Bliesner/LabControl_GMPEnvironment

Table 12.1 The laboratory control system sub elements and some example regulatory citations.

Laboratory control system sub element	Description	Some regulatory citations related to the sub element
Laboratory Managerial and Administrative Systems (MS)	The Laboratory Managerial and Administrative Systems sub element provides the infrastructure for efficient and compliant operations of an analytical laboratory. The sub element includes at least six individual topics. These topics are as follows: (i) organizational structure and roles and responsibilities, (ii) training and qualification, (iii) laboratory budgeting, purchasing, and requisition, (iv) laboratory administration and operations, and (v) laboratory chemicals, solutions.	**WL: 320-13-20** "You failed to establish an effective corporate and local system for managing quality which would include the appropriate organizational structure, procedures, processes and resources, as well as activities to ensure confidence that all APIs produced by your facility will meet the intended specifications for quality and purity." **WL: 320-13-21** "Per your training procedures, these questionnaire forms must be completed following each training to assess the individual's competence. The inspection documented over 40 instances of incomplete training records for three of your staff members." **WL: 320-16-28** "You did not provide details regarding your planned repairs and replacements, such as purchase orders and photographs of the renovations and replacements." **WL: 320-13-22** "Your firm's response mentions that the QC operations are now under "direct control of administrator", but it does not define the roles and responsibilities of the administrator to ensure the integrity and reliability of all QC laboratory data." **WL FLA-17-03** "Furthermore, you have not validated your reverse osmosis purified water system to demonstrate that you can effectively control, maintain, sanitize, and monitor the system so it consistently produces pharmaceutical grade water that, at a minimum, meets the USP monograph for purified water." **Form 483 Observation FEIN 3012333115, dated 11/08/17-11/17/17** "The critical reagent xxxx antibody used for drug substance identity testing by Western Blot was not verified for its specificity to demonstrate that the antibody only recognizes yyyy and does not cross-react with other serotypes. The drug substance identity is one of the critical product quality attributes tested at release." **WL # 557722** "For example, using previously received lots of raw ingredients as FT-IR reference standards, instead of using qualified compendial or non-compendial reference standard material, to verify the identity of raw ingredients would not be considered a scientifically valid method."

Laboratory Documentation Practices and Standard Operating Procedures (OP)	The Laboratory Documentation Practices and Standard Operating Procedures sub element forms a critical component of the overall Laboratory Control System (LCS). Much of the essence of CGMPs involves the generation, review, approval, revision, and control of procedures and data. The Laboratory Documentation Practices and Standard Operating Procedures sub element mirrors the LCS itself. Therefore, there are ten individual parts included in the sub element. These parts are as follows: (i) SOPs Related to the Laboratory Managerial and Administrative Controls Sub Element, (ii) SOPs Related to the Laboratory Documentation Practices and Standard Operating Procedures Sub Element, (iii) SOPs Related to the Laboratory Equipment Sub Element, (iv) SOPs Related to the Laboratory Facilities Sub Element, (v) SOPs Related to the Method Validation and Method Transfer Sub Element, (vi) SOPs Related to the Laboratory Computers Sub Element, (vii) SOPs Related to the Laboratory Investigations Sub Element, (viii) SOPs Related to the Data Governance and Data Integrity Sub Element, (ix) SOPs Related to the Stability Program Sub Element, and (x) SOPs Related to the General Laboratory Compliance Practices Sub Element.	From the good manufacturing practices (GMPs) Section 211.160 General requirements. "(b) Laboratory controls shall include the establishment of scientifically sound and appropriate specifications, standards, sampling plans, and test procedures designed to assure that components, drug product containers, closures, in-process materials, labeling, and drug products conform to appropriate standards of identity, strength, quality, and purity." **WL: 320-13-21** "In addition, address the root cause of your quality unit's failure to control and detect the manipulation or alteration of laboratory documents, and describe actions to prevent recurrence." **WL: 320-14-08:** "Appropriate record retention policies should also be in place. Our inspection revealed that your firm destroyed CGMP records directly related to the testing and manufacturing of your products. Your firm should reevaluate your record retention policy for all of your CGMP records." **Warning Letter 320-19-01:** "Our investigator observed a quality control analyst and laboratory team leader signing and backdating a test record." **WL MARCS-CMS-572991:** "Your notebooks and worksheets for the (b)(4) analysis of (b)(4) Brd Spect SPF 50 lack documentation of lot numbers and expiration dates of standards."

(Continued)

Table 12.1 (Continued)

Laboratory control system sub element	Description	Some regulatory citations related to the sub element
Laboratory Equipment (LE)	Since the vast majority of data in a modern analytical laboratory are acquired via instrumentation, it is essential that these instruments have been properly installed and qualified and are kept in proper working order. The Laboratory Equipment sub element addresses these points. Typically, the Laboratory Equipment sub element includes at least three individual topics. These topics are as follows: (i) Administrative Procedures Related to the Laboratory Equipment Sub Element, (ii) Equipment Specific Standard Operating Procedures (e.g. procedures related to qualification, operation, maintenance, repair, and calibration of each individual type of instrument), and (iii) Laboratory Glassware.	**WL: 320-13-22** "Deviations pertaining to laboratory equipment failures were not investigated. During the review of the service report log books for HPLC and GC units, the investigator found many instances of servicing due to instrument problems that were not documented as deviations. As such, your firm failed to follow the SOP 1019 entitled "SOP for Deviation Management." According to this SOP, all service activities for equipment, including laboratory instruments, need to be documented as deviations. Your response stated that the SOP has been changed to require deviations only for instances in which servicing was required to repair a problem with the instrument. Your response failed to address why no deviation was filed and investigated for the instances in which instrument problems were the cause of system maintenance (such as "system problem visit" on May 8, 2012, for Instrument QC/INST/067 or "needle motor stalled" on May 13, 2012, for Instrument QC/INST/022). Please also note, as a general laboratory practice, any equipment malfunction that may have an impact on quality control testing should be appropriately recorded and investigated." **WL 320-18-39:** ". . . . For example, the review of your laboratory system should include, but not be limited to, the suitability of all laboratory equipment, a fully remediated calibration program . . . ." **WL Case# 540675:** "In response to this letter, provide a detailed plan for qualifying, maintaining, and monitoring all laboratory equipment." **WL CMS # 568173:** "As part of your detailed CAPA, include an update on your new laboratory operations, including a timeline for qualification and transfer of testing onto new instruments. Also include a list of all new equipment and existing equipment that will be used in the new laboratory." **WL 320-18-38:** "In your response, you stated that dirty glassware used in all prior preparations and tests was the root cause for the failures; you claimed that rinsing the glassware before testing resolved the problem. Your response was inadequate because you have no scientific justification for the assigned root cause, nor have you implemented adequate corrective actions and preventive actions (CAPA)."

Laboratory Facilities (LF)	As with a manufacturing area, it is essential that the laboratory be properly designed and equipped. This ensures that the data generated accurately reflect the true composition of the sample from which they came. In addition, it insures a safe working environment for laboratory personnel. The Laboratory Facilities sub element includes four individual topics. These topics are as follows: (i) Physical Construction and Layout of the Laboratory, (ii) Utilities and Services, (iii) Laboratory Facility Upkeep and Maintenance, and (iv) Safety and Environmental Concerns.	**WL 14-NWJ-02:** "2. Your firm failed to have an adequate laboratory facility available for the testing and approval (or rejection) of components, drug product containers, closures, packaging materials, in-process materials, and drug products (21 CFR 211.22(b))." **WL 320-17-14:** "During the inspection, your facility's room temperature was warm and humid, requiring you to open the windows in an attempt to lower the temperature."

(Continued)

Table 12.1 (Continued)

Laboratory control system sub element	Description	Some regulatory citations related to the sub element
Method Validation and Method Transfer (MV)	Method validation is the process of demonstrating that analytical procedures are suitable for their intended use. The method validation process for analytical procedures begins with the planned and systematic collection of validation data to support analytical procedures. In addition, the method validation process is considered to be complete once the method has been successfully transferred from the originating to the receiving, or end user, laboratory. The Method Validation and Method Transfer sub element includes at least five individual topics. These topics are as follows: (i) Regulatory Requirements and Industry Guidance Related to Analytical Method Validation, (ii) Analytical Method Development, (iii) Analytical Method Validation, (iv) Verification of Compendial Procedures, and (v) Analytical Method Transfer.	**WL: 320-13-23:** "Your laboratory uses non-validated assays for testing of key starting materials and for an intermediate used in the manufacture of (b)(4) API. Please submit method validation reports to support the scientific validity of these methods. In addition, please submit the validation status of all other non-compendial analytical methods used in your laboratory along with a timeline for completing any necessary validation activities." **WL: 320-16-04:** "4. Your firm failed to establish and document the accuracy, sensitivity, specificity and reproducibility of test methods employed by the firm (21 C.F.R. 211.165(e)). Analytical method validation demonstrates that a testing procedure is suitable for its intended use. The outcomes of method validation are important to assess the quality, reliability and consistency of analytical results." **WL CMS# 558815:** "In your response, you stated that you would perform method validations for the assays specifically cited on the Form FDA 483. Your response was incomplete because it did not address the need to ensure that all your finished product tests and raw material methods are validated (or verified if compendial)." **WL Case # 562365:** "Upon our review of your most recent corrective actions, you still have not satisfactorily demonstrated that your UV spectrophotometer test methods are validated." **Form 483 Observation date 12/4/2017-12/12/2017 FEIN Number 3005531475:** "Specifically, the analytical test methods are not adequately transferred from the validating laboratories to this firm to assure that drug products meet appropriate standards of identity, strength, quality, and purity."

Laboratory Computer Systems (LC)	In a modern analytical laboratory, nearly every instrument is either controlled by, or has data residing and processed by a computer. Subsequently, it is critical that reliability of these systems and their associated software be evaluated. In addition, establishment, operation and maintenance of a Laboratory Information Technology (IT) infrastructure, functionality, and support system are an important component of the Laboratory Computers sub element. The Laboratory Computer Systems sub element includes at least four individual topics. These topics are as follows: (i) Laboratory IT System Administration and Procedures, (ii) Laboratory Computer Hardware, (iii) Laboratory Software, and (iv) Laboratory IT Related Services.	**WL MARCS-CMS-572991:** "You also did not have software validation for the Microsoft Excel spreadsheets used to calculate results for your modafinil and armodafinil assays as well as (b)(4) moisture content analysis for triamcinolone." **WL: 320-15-10:** "In your response to this letter, provide a comprehensive corrective action plan for computer system controls over all laboratory and manufacturing instrumentation and equipment. This response should include but not be limited to: • Information regarding changes in the reliability of your information technology infrastructure, including but not limited to improved computer systems, systems validation, revised procedures, and appropriate retraining of employees that will be implemented immediately to ensure your firm creates and retains complete and accurate electronic raw data."

(Continued)

Table 12.1 (Continued)

Laboratory control system sub element	Description	Some regulatory citations related to the sub element
Laboratory Investigations (LI)	CGMPs exist to not only prevent problems from occurring, but to mitigate and address deficiencies when they do occur. The Laboratory Investigations sub element is designed to systematically identify the root cause of failures, determine how to correct the problems, and to prevent them from reoccurring in the future. The Laboratory Investigations sub element generally includes (i) Contents of the Out-of-Specification Investigation SOP, (ii) Executing Laboratory Investigations, (iii) Documenting Laboratory Investigations, and (iv) Corrective and Preventive Actions.	**WL 320-18-62:** "Your firm invalidated out-of-specification (OOS) results without adequate investigation and scientific justification. Examples include: In March, 2017, you obtained OOS results for the (b)(4) impurity during stability testing of (b)(4) injection batches (b)(4). You suspected the analyst may have incorrectly rinsed the HPLC vials. New samples prepared and tested by a second analyst using both the original column and a new column, as well as old and new vials, also yielded OOS results. Although you lacked sufficient evidence, your investigation concluded that the OOS results were due to sample vial contamination. You invalidated the OOS results after obtaining passing results from testing retain samples." **WL CMS# 55206:** "*Assay failures* Your contract laboratory reported low failing assays for **(b)(4)** active ingredients in your OTC drug product **(b)(4)** SPF 30 **(b)(4)** lot **(b)(4)**. The OOS final investigation report summary stated that the low OOS sunscreen actives were confirmed through retesting, and there was no indication of laboratory error. The investigation was approved by QA personnel. No further assessment was performed, and the investigation lacked corrective actions. Despite this OOS, your firm distributed lot **(b)(4)** to U.S. customers." **Warning Letter 320-18-69:** "Your response is inadequate. You attributed this failure to an "unknown lab error." You claimed that the low individual assay test result was an outlier and that the most probable root cause was analytical error. Outlier tests have no applicability in cases where the variability in the product is what is being assessed, such as for (b)(4). You did not provide sufficient justification for disregarding the low result or supporting your unspecific conclusion of unknown laboratory root cause." **WL CMS # 557903:** "C. Your firm opened multiple manufacturing investigation reports and trending assessments from July 2016 to October 2017 related to out-of-trend and OOS content uniformity results for metolazone 2.5 mg tablets. A scientifically justified root cause had not been identified, and effective CAPA plans had not been implemented. Despite substantial non-uniformity observed in multiple batches of metolazone 2.5 mg tablets, you continued to manufacture and release this drug product up to the time of the inspection."

Laboratory Data Governance and Data Integrity (DI)	Data integrity is defined as the "…completeness, consistency, and accuracy of data. Complete, consistent, and accurate data should be attributable, legible, contemporaneously recorded, original or a true copy, and accurate (ALCOA)."[a)] Data governance, on the other hand, is best defined as the sum total of arrangements that provide assurance of data integrity. These arrangements ensure that data, irrespective of the process, format or technology in which it is generated, recorded, processed, retained, retrieved, and used will ensure a complete, consistent, and accurate record throughout the data lifecycle.[b)] Ensuring the integrity of data is a critically important component of the CGMPs and the pharmaceutical industry's responsibility to ensure the drugs they manufacture are safe, effective and have the quality they purport and of FDA's ability to protect public health. A properly constructed and operating data governance system will ensure that all data and records are integral.	**WL: 320-13-20:** "d. Unacceptable practices in the management of electronic data were also noted. The management of electronic data permitted unauthorized changes, as digital computer folders and files could be easily altered or deleted." **WL CHI-3-17:** "Our investigators observed that information technology (IT) staff at your facility share usernames and passwords to access your electronic storage system for (b)(4) data." **WL: 320-16-02:** "a. Your HPLC systems are configured so that no passwords are required to log in. Credentials are unverified. Anyone who accesses the system can use software administrator privileges, which means that there is no electronic or procedural control to prevent manipulation of data. b. Your HPLC system had no access controls to prevent alteration or deletion of data. Furthermore, your HPLC software lacked an audit trail feature to document all activities related to the chromatographic analysis. Because of this failure, neither your quality unit nor your laboratory staff could demonstrate that HPLC records included complete and unaltered data. They were also unable to verify that there had been no alterations or deletions. c. One of your analysts stated that another, unknown individual had logged into the system using the analyst's credentials. This unknown individual performed injections and deletions without the analyst's knowledge." **WL: 320-16-07:** " • altering time and date settings of computerized equipment using the software administrator's access privilege • manipulating test integration parameters to obtain passing or desirable results • aborting on-going sample analyses • over-writing and deleting raw data files containing original results"

(Continued)

Table 12.1 (Continued)

Laboratory control system sub element	Description	Some regulatory citations related to the sub element
Stability Program (SB)	The purpose of stability testing is to provide evidence of how the quality of an API or drug product varies with time under the influence of a variety of environmental factors such as temperature, humidity, and light. A properly designed stability program involves many individual parts, which may include (i) Stability Program Overview and Work Flow Description, (ii) Stability Study Initiation (Requests and Protocol Generation), (iii) Stability Study Activation (Sample Set-Down), (iv) Time Zero Testing, Review and Reporting, (v) Sample Pulls, (vi) Sample Testing and Reporting of Results, (vii) Data Analysis, Reporting, and Shelf-Life Determination, (viii) Stability Program Administration, (ix) Additional Types of Stability Studies, and (x) Stability Chambers.	**WL: 320-06-03** "2. Your firm failed to establish and follow an adequate written stability testing program designed to assess the stability characteristics of drug products and to determine appropriate storage conditions and expiration dates in that: A. There is no assurance that stability sample test intervals for each attribute examined have been met to assure valid estimates of stability [21 CFR 211.166(a)(1)]." **WL 320-18-56:** "Your firm failed to conduct on-going stability testing of (b)(4) and (b)(4) at various time intervals, as specified in your stability program. For example, in September 2016, you failed to perform testing at the 12-month stability time interval for (b)(4) batch (b)(4). Further, you performed the scheduled 6-month stability testing of (b)(4) batch (b)(4) four months late (in March 2017)." **WL CMS 552896:** "You have not established a stability program, including written procedures." **WL 320-18-64:** "Your quality unit also failed to ensure that samples intended for stability studies are stored with controlled temperature and humidity. Your firm kept retain and stability samples of (b)(4) USP in a cabinet in the quality control laboratory without monitoring temperature and relative humidity."

General Laboratory Compliance Practices (CP)	General laboratory compliance practices are those day-to-day practices and activities which everyone working in a CGMP compliant laboratory are expected to know and understand, yet they might not be clearly described or codified in writing. These compliances practices address things like (i) Housekeeping and General Appearance, (ii) Sample Handling and Storage, (iii) Standard Handling and Storage, (iv) Reagent Handling and Storage, (v) Instruments and Equipment, (vi) Laboratory Water and Solvents and Solutions, (vii) Laboratory Facilities and Services, (viii) Laboratory Supplies, (ix) Laboratory Glassware, (x) Analyst Documentation Practices, (xi) Laboratory Safety – High Level Review, (xii) Analysts Demeanor and Behavior, and (xiii) Supervisor Presence and Effectiveness.	**WL: 320-13-21:** "b. On March 18, 2013, an FDA investigator identified the presence of unlabeled and partially labeled vials in the laboratory glassware washing area." **WL: 320-14-01:** "2. Identify and interview your current employees who were employed prior to, during, or immediately after the relevant period(s) to identify activities, systems, procedures, and management behaviors that may have resulted in or contributed to inaccurate data reporting." **WL 320-18-12:** " .... Elements of your CAPA should include .... It should also include improved laboratory supervision of analysts."

a) USFDA Draft Guidance to Industry titled "Data Integrity and Compliance With CGMP," December, 2018.
b) Pharmaceutical Inspection Convention Pharmaceutical Inspection Co-Operation Scheme (PIC/S) Guidance titled "Good Practices for Data Management and Integrity in Regulated GMP/GDP Environments," November 2018.

Some Final Thoughts on Establishing and Maintaining a Compliance Laboratory Control System

For a myriad of reasons, many pharmaceutical companies see the CGMP regulations as an imposition by Government and a burden to their organizations. They see no value in them and consider regulatory compliance as just another cost of doing business. In addition, they rarely consider the cost of non-compliance, in that it is very expensive in terms of time and effort to address lengthy Form 483 observations, Warning Letter citations, or, even worse, deal with the tremendous price of a Consent Decree.

This is unfortunate, because in the author's experience when the systems and principals presented in this text are implemented within QC and R&D laboratories, and in the company at large, the chances of withstanding regulatory scrutiny, enhancing operational efficiency and thus increasing profitability, improve significantly. So, in the end, experience shows that:

Following the CGMPs = Good Compliance = Good Science = Good Business

And that just makes sense.

Index

Laboratory Control System Operations in a GMP Environment, First Edition. David M. Bliesner.

Companion website: www.wiley.com/go/Bliesner/LabControl_GMPEnvironment